Paulo Alex Nakata
Milton Parron Padovan

Technologies for sustainable rural development

Paulo Alex Nakata
Milton Parron Padovan

Technologies for sustainable rural development

Methodologies for sharing with rural producers and assessment of socio-economic and environmental impacts

ScienciaScripts

Imprint

Any brand names and product names mentioned in this book are subject to trademark, brand or patent protection and are trademarks or registered trademarks of their respective holders. The use of brand names, product names, common names, trade names, product descriptions etc. even without a particular marking in this work is in no way to be construed to mean that such names may be regarded as unrestricted in respect of trademark and brand protection legislation and could thus be used by anyone.

Cover image: www.ingimage.com

This book is a translation from the original published under ISBN 978-613-9-61491-2.

Publisher:
Sciencia Scripts
is a trademark of
Dodo Books Indian Ocean Ltd. and OmniScriptum S.R.L publishing group

120 High Road, East Finchley, London, N2 9ED, United Kingdom
Str. Armeneasca 28/1, office 1, Chisinau MD-2012, Republic of Moldova, Europe
Printed at: see last page
ISBN: 978-620-6-02583-2

Copyright © Paulo Alex Nakata, Milton Parron Padovan
Copyright © 2023 Dodo Books Indian Ocean Ltd. and OmniScriptum S.R.L publishing group

BIOGRAPHY

PAULO ALEX NAKATA, son of Tomoaki Nakata and Rosalinda Kenko Miyai, born on 28 May 1981, in Loanda - Paraná.

Joined the General Administration course in 2006, concluding in December 2010, at Universidade Nove de Julho - UNINOVE.

Joined the MBA Lato Sensu in 2010, concluding in December 2011, at the Law School Prof. Damásio de Jesus, FDDJ, Brazil.

He joined the Masters course of the Postgraduate Programme in Agribusiness at the Federal University of Grande Dourados in March 2013, defending his dissertation 28 in August 2015.

SUMMARY

GENERAL SUMMARY

During the period from January to November 2014, research actions were developed involving family farmers in the southern region of Mato Grosso do Sul. Thus, the objective was to know alternative means to achieve sustainable development based on the constructs, especially those advocated by the *Triple Bottom Line* whose main assumption says that Sustainable Development will occur from the balance between the purposes of economic, social and environmental dimensions through the evaluation of the impacts of spring/summer green manures preceding the cultivation of corn. A broad bibliographical review was carried out in order to list innovations of technology transfer, where the generation and exchange of knowledge and technologies are some of the great challenges for sustainable rural development. The first one deals with **"Innovative methodologies or processes of technology sharing with technicians and rural producers"**, whose objective was to make a relevant literature review in order to list methodologies that have an innovative character in relation to those predominantly used in different regions of Brazil, where the consolidation of existing networks and the application of innovative forms and models of exchange and collective construction of knowledge have proven to be an essential approach for sustainable development. The second chapter deals with the theme: **"Socioeconomic and environmental impacts of the cultivation of green manures prior to corn cultivation under agro-ecological management"**. The research that underlies this chapter was carried out in the central southern region of the state of Mato Grosso do Sul; the data were obtained by interviewing nineteen rural producers, users of the technology after the application of the *Snowball Sampling* methodology, and were carried out using a script according to AMBITEC- SOCIAL and AMBITEC- AGRO. A complementary evaluation was also carried out with the interviewees, by means of questions prepared by researchers from Embrapa Agropecuária Oeste. The results showed positive social impacts, such as the increase in income generation, as for the environmental ones, they refer to the drastic reduction in the use of agrochemicals/inputs, as for the economic impacts, the results pointed out the economic viability under the Net Present Value point of view.

Key words: Technological innovation, green manure, economic viability, sustainable development, environmental and personal health.

GENERAL INTRODUCTION

Humanity has gone through several eras and, with them, several were man's discoveries and evolutions. For example, hunting for our own food, creation and domestication of animals, invention of the wheel and various means of transport, and industrialisation represent innovations in search of better living conditions for people. Nowadays, we live in the information age. This new era has changed the whole society and created a new economy, which is based on the principle of the ability to accumulate knowledge and transform it into inputs for new products and knowledge, creating a productive circle for organizations and for society (NETO et. al., 2012).

However, for this dialogue to occur in a systematic and effective way, it is necessary to adopt methodological tools that allow prospecting the demands of the different productive sectors and, in the opposite direction, that also allow evaluating the impact resulting from the use of technological solutions transferred or shared, feeding back the process of research and development (GASTAL, 2013). Also for the author, each mode of development also has a structurally determined development principle that serves as a basis for organizing technological processes. For example, industrialism is focused on the development of the economy, that is, to maximize production. On the other hand, informationalism aims at technological development, that is, the accumulation of knowledge and higher levels of complexity in the information process, transforming them so that the target audience can understand them, especially in rural areas, where the predominant level of education is still low (NETO et. al., 2012; RAI et al., 2013).

In this way, technology sharing can be considered as a process that lies within informationalism, whereby a technology provider communicates and transmits the technology through multiple activities and forms to recipients, helping them to manage the challenges of using this information to create change within their work settings, involving strategic efforts to disseminate innovative information and scientific practices to individuals, organisations and communities (LEE et al., 2010; RAI et al., 2013).

Perera (2009) points out that, in general, the technologies generated achieve low rates of adoption by the target audience, that is, in this case, farmers. Perera (2009, p. 12) emphasizes that this gap between the generation of technologies by research centres and their acceptance by their target audience can be given by variables ranging from the [...] "communication model adopted for dissemination to the inadequacy of the technologies generated, caused by researchers' lack of perception of the reality of farmers and their problems and anxieties" increasing the distance between them.

Thus, contributing to the achievement of advances in processes of technology sharing and the increase of its adoption and incorporation in agricultural production arrangements, also measuring some impacts arising from the adoption of technology, becomes the overall objective of this research, justified by the need to expand the perception in enabling the incorporation of technologies in agricultural production systems.

It is important to emphasize that it is not enough just to develop technologies for the agricultural sector, but also to expand the strategies and methodologies that contribute to the incorporation of these technologies in agricultural production systems (PADOVAN et al., 2011).

In this context, bibliographical and field research work was carried out, the results of which make up the scope of this dissertation, which consists of two chapters, the first being: Sharing technologies with technicians and rural producers through innovative methodologies or processes and the second: Socio-economic and environmental impacts of growing green manures prior to maize under agro-ecological management.

CHAPTER 1

INNOVATIVE METHODOLOGIES OR PROCESSES FOR SHARING TECHNOLOGIES WITH TECHNICIANS AND RURAL PRODUCERS

Summary
The purpose of this research was to survey a broad literature review so as to identify methodologies that are innovative in relation to those predominantly used, with views to sharing technologies throughout the Brazilian territory. The main results refer to the identification of the main methodologies of technology sharing currently used in different regions of Brazil, where the consolidation of existing networks and the application of innovative forms and models of knowledge exchange and collective construction of knowledge proved to be an essential approach for the development of family farming in Brazilian rural identity territories.

Keywords: Technology transfer methodology, technological innovation, socioeconomic development.

Abstract
The research aimed to the lifting of a broad literature review in order to cite methodologies that have innovative character vis-à-vis the predominantly used to technology transfer throughout the Brazilian territory. Where the main results refer to the identification of the main technology transfer methodologies currently applied to different regions of Brazil, where the consolidation of existing networks and the application of innovative models and forms of exchange of knowledge and collective construction of knowledge proved an essential approach to the development of family agriculture in territories of rural Brazilian identity.

Keywords: Technology transfer methodology, technological innovation, socioeconomic development.

1 Introduction

The generation and exchange of knowledge and technologies are some of the major challenges for sustainable rural development, where the entry of new actors and technologies into the market, when combined with new economic and demographic pressures, suggests the need for more innovative and less linear approaches to promote a technological transformation in different sectors of society (SPIELMAN et al., 2009; BALSADI et al., 2013).

Thus, the adoption of innovative technologies and the use of new knowledge can modify the productive conditions and quality of life of farmers and their families through processes and knowledge generated by research in Brazil (PADOVAN et al., 2013).

However, Cereda and Vilpoux (2010) point out that many technologies made available by research are not always adopted, generating a stock of technologies. In this context, a strategy to increase the adoption of technologies would be to improve the processes of technology transfer or sharing and make the processes of building new knowledge more efficient, with multidisciplinary teams and methodologies that adapt to the needs of farmers, in addition to strengthening organizational networks (BALSADI et al., 2013).

Efforts on undertakings in the generation and exchange of technological knowledge and its availability in different social contexts, need to be supported effectively, at different levels of technical assistance and rural extension, supported by appropriate methodologies for sharing technologies in order to provide sufficient capillarity to reach and meet the different ecoregions of the country (SPIELMAN et al., 2009; BALSADI et al., 2013; PADOVAN et al., 2013).

Thus, it is necessary to have a more comprehensive view of the production chains and the main processes involved, considering issues that ensure greater interaction with the market (economic aspects), i.e., far beyond the focus of the predominant performance of agricultural research that focuses predominantly only on the problems faced from the "gate inwards", as stated Balsadi (et al, 2013). From this perspective, the question that arises is: "are there innovative methodologies for sharing technologies capable of facilitating technological adoption by rural producers"?

The need to identify methodologies for sharing technologies that are considered innovative in relation to those predominantly used is becoming increasingly evident. In this context, we also seek transformation through innovation, overcoming obstacles and adversities, transforming them into great opportunities for sustainable development, employment and income generation, food security and local development.

1.1 Innovation

Innovation originates from the Latin - *innovare* - which means to do something new or to introduce a novelty. However, the innovative activity in organizations involves, in a simple way, the organizational direction for using opportunities with new practices and processes that may arise in the market (SILVA et al., 2012).

This understanding is better understood as from the moment that innovation can be considered as the result of a process of resources and competences combinations to generate inventions and make them available to the market (ZEN, 2007). In other words, the ideas that arise should be transformed into opportunities or into something practical of extensive use, bringing changes to the organization and its environment, successfully integrating a new idea or product in a process that includes technical, economic and social components (PLONSKI, 2005; SARKAR, 2008; SPIELMAN et. al., 2009).

In this context, three important characteristics stand out. First, innovation is the creative use of different types of knowledge in response to existing social and economic needs and opportunities. Secondly, an idea only becomes an innovation when it is adopted as an integral part of a process, where many agents try new things, but little of these experiments produce practices or products that improve what is already in use. Third, it is based on complexity theory, where innovation should not only result from something new, but also in the search for improvements in the processes currently used, incorporating long and complex processes (SPIELMAN, 2009).

However, to transform technological knowledge into innovation, it is necessary that these are shared with audiences that can incorporate them into their production processes, promoting the breaking of existing cultural ties of using conventional methodologies used for decades and incorporate other methodologies that represent innovations in the processes of sharing technologies with technicians and rural producers (PADOVAN et. al., 2013; SILVA, 2013).

2. Innovative methodologies for sharing technologies

The basis for any type of development is the capacity of individuals, organisations and society to improve what is currently being done. Thus, to discuss methodologies for sharing knowledge and technologies in order to promote sustainable rural development (SRD), one assumption is basic: the technology should be appropriate to the reality of those who receive it (BALSADI et al., 2013; GASTAL, 2013).

It is essential that efforts to generate and share knowledge and technologies and their availability to different social contexts need to be effectively supported by a robust network of professionals dedicated to providing technical assistance and rural extension (BALSADI et al., 2013). However, the authors emphasize that the multiplication does not occur only from the professionals already engaged in the labour market, but also by the adoption of methodologies for sharing innovative technologies that need to expand [...] "incorporating the concepts of market intelligence, marketing and social marketing, so that they are more strategic and ambitious in the process of sharing technologies".

Table 1 summarises all the methodologies listed in this work.

Table 1 Objectives and innovative factors of knowledge sharing methodologies.

Methodology Year UF	Objective	Innovative factors
Training and Visit 96/97 PR T&V	Motivate and train private and public technical assistance and rural extension (ATER) agents.	Reorganization of rural extension agencies, feeding research programmes with problem-solving found.
Capacity Building 2010 MT continued	Motivate and train technical assistance and rural extension agents (ATER) with a focus on milk production	Fulfilling unmet needs contemplated in the T&V methodology
2011 MS Unity , ReferenceMT , Technological-AM URT	Reproductionsystem production of grains, fibre, meat, milk, wood products in intercropping, in succession or crop rotation on a reduced scale - iLPF.	Encourages the adoption of new technologies, attitudes and behaviours, a fact that implies changes in the vision of rural producers in relation to production.
Units-2007 MS	Dairy production processes, agro-	It provides opportunities for training

Reference - UR	ecological production arrangements , recuperation of riparian vegetation in spring areas and beekeeping	and interaction with innovative producers, maintaining the essence and peculiarities of each UR, facilitating communication between farmers.
Research2011 MG Participatory" with and for farmers"	It seeks a new type of research that restores the value of farmers as subjects of their development.	It made it possible to achieve production by the masses and not mass production, which is the essence of a social inclusion process.
Peasant a2008 SE Peasant	Seeks to exchange basic information agro-ecological.	It has made possible the irradiation of agroecological knowledge and the plot of the future.
Pedagogy 1969 BH Alternation	Increasing the capacity of regions to formulate development strategies through local productive clusters and integration with local actors	It seeks sustainable alternatives, based on care to guarantee the present and the lasting future and centred on the development of the people in the field.
Field Day at al TV	naNacion Sharing of technical information on highly relevant topics, using language that is more appropriate and realistic predominant among farmers	The potentialities opened up by new interactive technologies, such as the internet and social networks, offer a reference challenging and innovative more adaptable for young people
ProseNE	Rural Using the airwaves like radio to share technical information and knowledge that contributes to improving the quality of life of people in the countryside	Extremely efficient in reaching the public most in need of information, as well as the need for better using time in the age of computerisation
Intervention2007 MS, PE, PA	collectiva Developing arrangements interacting with the reality of family farmers using its own dynamics.	They prioritise social technologies, enabling solutions for problems in the field with the interaction of technical and popular knowledge.

Source: Prepared by the author.

Following are briefly presented and discussed all methodologies for sharing technologies and knowledge presented in Table 1, which are being qualified as "innovative" and can contribute to increase innovations in Brazilian agriculture and cattle ranching.

2.1 Training and Visiting

The Training and Visiting (T&V) methodology was developed by Daniel Benor and James Q. Harrison in 1977, initially used by the World Bank, which financed and disseminated it mainly in Africa and Asia (MARTINS; GALERANI, 2007). It was adopted and modified for the conditions of the country by Embrapa and Emater in Paraná, with great success, on the occasion of the 1996/1997 Crop Plan (DOMIT, 2007). The author points out that the T&V methodology motivates and enables technology transfer agents from research and official and private technical assistance and rural extension (ATER) to establish regional partnerships with a view to developing future joint projects.

The basis of T&V is the identification of the actors of the diffusion subsystems functioning through the continuous training of groups of agents of technical assistance and rural extension (ATER), official and private called Multipliers I (TMI), which remain in direct contact with researchers and other specialists, being informed of the technologies available in research institutions, being responsible for the transfer of such knowledge to field technicians, called Multipliers II. The (TMII), perform the function of passing on the technologies on their properties (MARTINS; GALERANI, 2007; OLIVEIRA; LIMA, 2007; ALVES; MODESTO JÚNIOR, 2013), as shown in Figure 1.

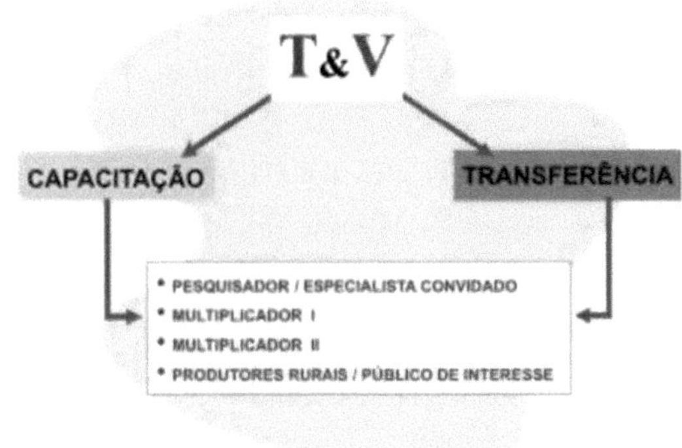

Figure 1. Schematic representation of the T&V methodology and groups involved.
Source: Domit (2007).

The methodology goes beyond the issue of dissemination of information and suggests, as a fundamental condition, the reorganization of rural extension agencies, thus becoming a method of feedback of research programmes, enabling the return of problems and difficulties encountered in the application by producers, fundamental requirements of this methodology (MARTINS; GALERANI, 2007). The authors also advocate that one of the great advantages of the method is its multiplier effect. However, if producers differentiate themselves over time, they cease to be a reference. In order to avoid this differentiation, the exchange of producer groups directly assisted by the TA agent should be recommended, so that all can benefit from the system equally, which is unlikely in the Brazilian model of T&V.

Therefore, T&V can be defined as a methodology that ensures the flow and control of technological information, creating the "responsibility" of the feedback of the difficulties encountered by producers when applying a technology in their production system with an efficient technology transfer management system that enables the organised implementation of already existing extension principles.

According to Martins and Galerani (2007), because of its versatility, the T&V enables it to be adopted by various areas of knowledge, but it depends on good management and the dedication of its components to make the process of sharing technologies more agile, directly influencing the increase in income, improvement of preservation and the productive environment of farmers assisted in projects that use the methodology.

2.2 Continued training

Continued training can be considered as a permanent process of improving the knowledge required for professional activity, carried out after the initial training, in order to ensure better quality knowledge to technicians and then to farmers, since the need for knowledge, technologies and new market demands imposed by innovation favor the process of sharing technological knowledge (FRIGOTTO, 2011).

The continuous training methodology was applied in 2010 in the "Continuous Training in Dairy Cattle Raising" project for technicians in the state of Mato Grosso. The objective was to train, guide and empower ATER technicians to form a group of 50 technical advisors specialised in dairy farming with quality to guide and train other field technicians, producers and others involved in this production chain (DOMIT, 2013), supplying the needs not covered in the T&V methodology.

Thus, the action strategy was developed in three stages: 1ª) Management - Formation of a Management Group composed of representatives of partner institutions and the coordination of Embrapa; 2) Responsibility for articulation, organization, monitoring and evaluation of performance and alignment necessary for the success of the project and 3) Definition of the number of technicians involved (DOMIT, 2013).

The training of multiplying technicians was the second stage, which was divided into five modules: 1) Technical levelling and information on the project; 2) Bulk production; 3) Integrated milk production system; 4) Development of sanitary programmes and 5) Animal reproduction and improvement and strategic management of rural properties.

In the third stage, another technology sharing methodology was added called "Technological Reference Units" (URTs) to promote the dissemination of practical instructions through units installed in the field. The URTs were implemented through partnerships with local technical assistance entities, municipal/state agriculture secretariats or similar, and also previously articulated producers.

According to Domit (2013), in the URTs when installed in rural properties, farmers must commit to conduct them according to previously agreed techniques and make them available for visits, field days and practical classes that may be necessary. The URTs are conducted by trained technicians and by the farmers responsible, who are committed to collecting and analyzing data, applying technologies, sharing them and disseminating results.

The fourth and final stage of the ongoing capacity building process is called "Communication and Technology Transfer", through which a communication channel is established between the researcher, the technician and the producer with a view to improving the flow of information between the various actors involved (DOMIT, 2013).

According to Lopes (2011), from modules 1 and 2, the evaluation data were compiled, analysed and compared to the introductory modules, and the technicians were grouped by public or private institutions" (DOMIT, p. 326, 2013), as can be seen in figure 2.

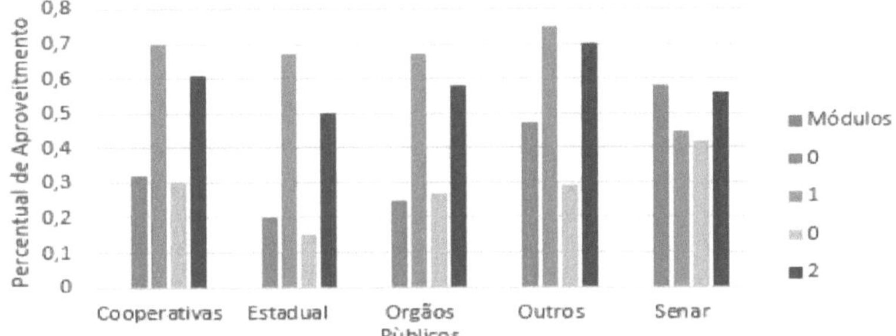

Figure 2 - Graphical representation of the performance evolution based on the technicians' evaluations according to training modules 1, 2 and 3 and institutional link.
Source: Domit (2013, p. 327).

Also according to Domit (2013, p. 326), "the evolution of the technicians in both modules reached 220%". It was also observed that the public servants showed a slightly lower performance than the private initiative professionals before the training, reflecting the "(...) unsatisfactory current situation of the official ATER network in the state of Mato Grosso" (DOMIT, 2013, p. 326).

Therefore, the "continued training" methodology provides improvements in the knowledge levels of technicians working in different production chains, providing better use in technology sharing processes. It is important to note that the structural basis of continued training is the T&V methodology.

However, some innovative factors are added, supplying needs not contemplated in the T&V methodology (DOMIT, 2013).

2.3 Technological Reference Unit

The Technological Reference Unit (URT) is a variant of some methods already used for technology sharing, developing various practices with a view to disseminating concepts capable of inducing the development of productive strategies adapted to the conditions of each rural property. This methodology has been experienced from systems of Crop-Livestock-Forest integration (iLPF) (BALBINO et al., 2011). The authors emphasize that the iLPF URTs comprise physical models of production system implemented in public or private areas, whose purpose is the reproduction of a productive system of grain, fiber, meat, milk, timber and non-timber products, among others, performed in the same area, in intercropping, in succession or rotation, but on a reduced scale. The authors also point out that there is a technological roadmap previously established for the implementation and conduct of the Technological Reference Units.

These production arrangements with iLPF TRUs make it possible to validate and demonstrate the maximization of the use of the biological cycles of plants, animals and their respective residues, as well as residual corrective effects and nutrients, inducing the development of appropriate production strategies, based on principles that favor the financial viability of rural properties and conservation and improvement of natural resources (DE ALMEIDA et al., 2012). The authors pointed out that in Brazil there were 186 Technological Reference Units in 2012 with iLPF, implemented by Embrapa and its partners, to validate and share technologies. These units were distributed in the Brazilian biomes: Amazonia (22), Caatinga (16), Cerrado (34), Atlantic Forest (103), Pampas (7), Pantanal (1) and Tabuleiros Costeiros (3).

In the state of Mato Grosso do Sul, in 2011 there were 4 URTs in iLPF (BALBINO et al., 2011), where the effort employed in technology sharing activities, research and training of technicians is reflected in the increasing adoption of integrated production systems, which are being supported by public policies and are estimated at 9 million hectares by 2020 (BALBINO et al., 2011; NICODEMO; MELOTTO, 2013).

According to Almeida et al. (2012), by establishing examples of the operation of production systems and technologies best suited to local conditions, the URT favours the adoption of new technologies, attitudes and behaviours, a fact that implies changes in the vision of farmers and their relationship with production, presenting great potential for development providing greater income diversification and, mainly, resulting in changes regarding the management of rural property.

2.4 Reference Units

The Reference-Unit (UR) methodology is a way to perform the sharing of technologies and knowledge using production units of innovative producers, which can serve as "reference" for others who have similar characteristics (PADOVAN et al., 2011a; PADOVAN et al., 2011b; PADOVAN et al., 2013). The authors emphasize that this technology sharing methodology starts from the "reality of the rural producer" and his family, without interfering to improve the visual aspect of the rural property, for example, respecting the producer's culture and expectation, his suitability and ability to interact with others.

Padovan et al. (2011a) emphasise that the Reference Units can contemplate from very incipient innovation initiatives up to high levels of innovation, that is, situations that represent a great diversity of rural producers. The URs serve as a base for the development of collective activities, such as: interactive technical visits, field days, courses, practical workshops, among others, for a wide range of audiences that are at different stages of technology adoption.

For a particular farmer's crop arrangement, production system or even a management process to become a "Reference Unit", the technical team working in that location interacts with this "actor", identifies his predisposition to share his innovation with others, supports him technically and methodologically to advance further, provides opportunities for training and interaction with other innovative farmers, and equips him to accelerate his innovation and incorporate other innovations. However, great care should be taken not to "force" the insertion of a new technology or form of management in the UR, for example, if the farmer or his family does not fully want it and does not see it as important and easy to adopt (PADOVAN et al., 2011a, b; PADOVAN et al., 2013).

According to Padovan et al. (2011a), the adoption of URs is a simple, feasible, versatile and dynamic methodology for Embrapa and other Research and Development (R&D) institutions to innovate in sharing technologies for agriculture and livestock production, whether for small, medium or large-scale producers.

The "Reference Unit" methodology was tried out in the Grande Dourados Territory, in the state of Mato Grosso do Sul, during the period from 2007 to 2011, covering the following themes: processes focused on dairy production, production arrangements on an agro-ecological basis, recovery of riparian vegetation in spring areas and beekeeping (PADOVAN et al., 2011a, b; PADOVAN et al., 2013).

When collective activities are carried out in the URs, it can be seen that when farmers identify technological innovations in real field conditions, they are strongly influenced to believe that it is possible to adopt those innovations, inducing them to incorporate new technologies into their production systems, stimulating initiatives to implement similar systems, adapting them to their local conditions (PADOVAN et al., 2013).

Therefore, with the adoption of URs it is possible to realize the ease in sharing technologies, as well as the construction of knowledge, which accredits it as an innovative methodology, serving as a reference for research and development institutions, technical assistance and rural extension (PADOVAN et al., 2011a,b; PADOVAN et al., 2013).

Participatory research "with and for" farmers

Participatory research "with and for" family farming can be defined as [...] "a multiple and differentiated repertoire of experiences of collective creation of knowledge aimed at overcoming the opposition subject/object within processes that raise knowledge and in the sequence of actions that aspire to generate transformations" (BRANDÃO; STRECK, 2006, p. 12).

Through this methodology, research is initially sought, but as it is a collective process, with active participation of farmers, knowledge and technologies are permanently shared, and the various participants in this process are protagonists of all stages, introducing innovations or improvement in the productive or social environment, resulting in new products or services (DE NOVAES; GIL, 2009).

The actors involved participate from the "identification of needs or research problems and project design to the implementation of the search for alternatives for the solution, there being no ready-made models, but rather the collective construction of a model" of learning (GASTAL, 2013, p. 47), not only of farmers, but also of technicians and researchers (OLIVEIRA et al., 2009).

A representative example of this type of action was the Unaí project, in Minas Gerais, which began in 2011, and its actions were based on four complementary and inseparable lines, as shown in figure 3.

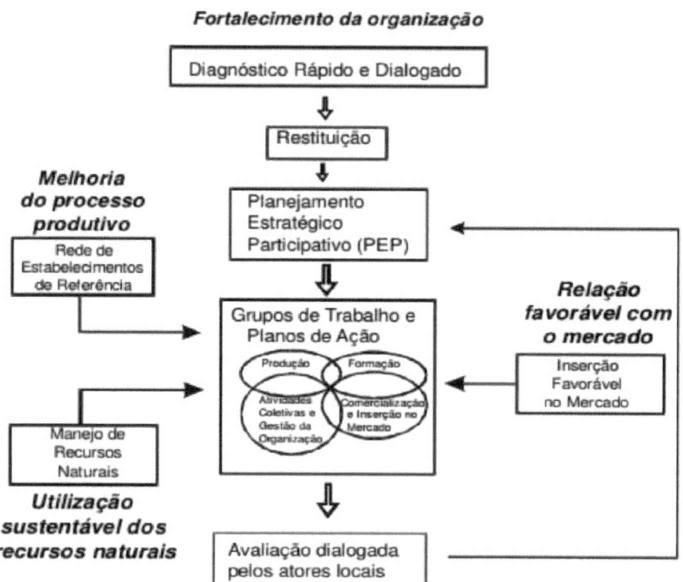

Schematic representation of the methodological device used in the Unaí Project.
Source: Gastal (2013), based on Sabourin et al. (2009).

Through this process, it was shown that to achieve social and productive inclusion of family farming and traditional communities, a new type of research was needed that incorporates the subjectivity of the human being from the dialogue between subjects, whose profile and methodologies adopted sought the interaction between economic and social interests in order to find alternatives for family farmers and settlers and thus rescue the appreciation of farmers as subjects of their development (BELTRÃO, 2013).

The project focused on specific problems based on the farmers' demand for appropriate solutions in social, ecological and economic terms, in addition to the learning process, of all those involved, representing technological solutions for social inclusion and improvement in the quality of life and the strengthening of social organisations (OLIVEIRA, 2009; GASTAL, 2013).

During the technological exchange process, the true validation occurs in 2013 of a technology understood as part of R&D, making clear the need for investment in farmers' reflection on the generation and adaptation of technologies to the reality of farmers in "their vision" and not the one idealised by researchers, scientists and technicians, since it is the farmers' sovereign decision to adopt or not a methodology, "making it possible to achieve production by the masses and not mass production, which is the essence of a social inclusion process" (GASTAL, 2013, p. 52).

2.6 Farmer to Farmer

The Farmer-to-Farmer methodology is a tool based on the exchange of agroecological information adopted in traditional communities in Central American countries, which aims to empower families. In this methodology, the farmers themselves are the protagonists of the process of building and sharing agroecological knowledge (HOLT GIMÉNEZ, 2008; SOSA, 2010).

This methodology has been adopted for about 20 years in countries such as Nicaragua, Cuba and Chile, enabling an environment of exchange of experiences in which farmers are the main protagonists: teaching, learning, exposing the challenges and potentialities of their farming systems to then operate possible technological innovations (LÒPEZ RODRIGUEZ, 2008; FONTES et al., 2013).

In Brazil, the methodology possibly emerged in Paraíba, with farmer experimenters, mainly in the Paraiba hinterland, and in Pernambuco, with farmer-multipliers, through the Centro Sabiá in the Zona da Mata and Agroflor in the hinterland (PIRES; SANTOS, 2007; PETERSONS; SIQUEIRA, 2007; SILVA et al., 2010). According to the authors, there are several successful experiences of traditional farmers with agro-ecological transition in which the Peasant-to-Peasant methodology played a central role.

In the period from 2008 to 2012, a research was conducted in the Sul Sergipano rural identity territory, located in the coastal tablelands of the state of Sergipe, with which it aimed to adjust the Peasant to Peasant methodology aiming at the diffusion of successional agroforestry in the conditions of northeastern Brazil (SIQUEIRA et al., 2013). The authors point out that in several collective activities carried out, it was sought to articulate a conversation circle that begins with the presentation of the participants and then a debate on the feasibility of a way to practice models of agriculture seeking to build an agroecological concept. Doubts were answered on visits to rural properties with innovative experiences, learning in practice about the type of management employed, the techniques used, among other novelties.

A total of ten exchanges were carried out and at the end of each exchange the evaluation of the experiences was promoted, which was guided by means of three questions: "what it takes away", "what it places", referring to some necessary adjustment of the experience and a final question "what it takes away", referring to the knowledge that was obtained (SIQUEIRA et al., 2013).

To assess the viability of this process, a survey was carried out in which a strong interaction with the Territorial Collegiate was perceived, enhancing agro-ecological innovation in the region. There was a greater perception of the need for the participation of the whole family in the processes of exchange, in addition to the consolidation of existing networks, enabling innovative ways of exchanging experiences, enhancing the sharing of knowledge and technologies (SIQUEIRA et al., 2013).

Therefore, the Farmer-to-Farmer methodology can be considered innovative for the purposes of technology sharing, as it has proved efficient for irradiating agroecological knowledge, in addition to the collective construction necessary for the introduction of new knowledge involving agroecosystems of greater complexity, such as successional agroforestry, called by Fontes et al. (2013) as the "Plantation of the Future", since they refer to agricultural production systems on an ecological basis, with great potential for promoting social inclusion and economic viability, especially for the most decapitalized farmers.

2.7 Pedagogy of Alternating Cycle

A small number of French farmers dissatisfied with the educational system of their country, which did not meet the specificities of an agrarian education for rural areas, initiated in 1935 a movement that culminated in the emergence of the alternating cycle pedagogy (GIMONET, 1999; MAGALHÃES, 2004; TEIXEIRA; BERNARTT; TRINDADE, 2008). The basic idea was to reconcile the studies with the rural activities of their families (TEIXEIRA; BERNARTT; TRINDADE, 2008).

In Brazil, the alternance pedagogy emerged in 1969, when the main objective was to act on the interests of the man in the field, mainly with regard to raising their cultural, social and economic level (TEIXEIRA; BERNARTT; TRINDADE, 2008).

The pedagogy of alternation can be defined as a teaching methodology, initially adopted by Family Centers of Formation through Alternating Cycle (CEFFAS). The methodology proposes alternating periods of formation, composed of one intensive week of formation in the Teaching Unit (Agro-technical School or University, for example, called School-Time) and two weeks in the property or social-professional environment, qualified as Community-Time, thus increasing the capacity of the regions to formulate development strategies through local productive arrangements and integration with local actors.

However, students take new technologies they have learned at schools, universities and research centres to rural communities and, in the communities, they also identify bottlenecks in production processes which serve as "inputs" for classes at the Teaching Units and for new studies at the Research Centres, thus building an important "two-way street" (SCANDOLARA et al., 2008; TEIXEIRA; BERNARTT; TRINDADE, 2008).

In this way, Rigamonti et al. (2013) emphasise that the strategies of sharing knowledge and technologies for family production units are dynamic and viable because they integrate various partners into the process, such as families, school, extension, research and community, facilitating the feedback of the system with the support of the students who act as "Rural Development Agents", developing activities as multipliers of the second generation in the process of sharing technologies and knowledge, one of the pillars of the alternating cycle pedagogy. In this process, the important role played by the farmer-experimenters who act as "anchors" of this work in the rural communities is emphasized. The authors also point out that this knowledge and technology are also developed in the production units and communities, with the support of young people and rural producers, integrating school, family, community, research and extension, facilitating the process of technological and social inclusion, as shown in figure 4.

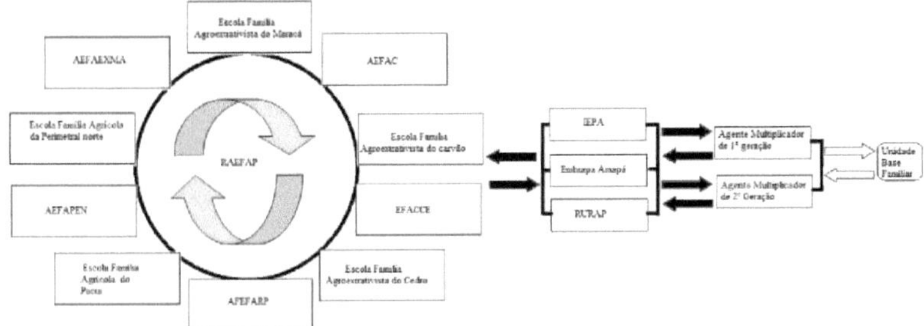

Figure 4. Flowchart representing the experience in the TT process in Amapa state.
Source: Prepared by the author based on Rigamonti et al. (2013).

In this context, the alternating cycle pedagogy "portrays the subject in the systemic process of socio-professional training, with epistemological and technological approaches, in which social actors integrate resources and feedback the systems, with a strong performance of the Agricultural Family Schools - EFAs" (RIGAMONTI et al., 2013, p. 312), supported by sustainable alternatives, based on care to guarantee the present and the lasting future and focused on the development of people in the countryside.

Therefore, according to Passador and Lopes (2014), the processes undertaken by farmers in the field also teach, generate and share technologies, this being one of the basic precepts of the Alternating Cycle Pedagogy.

2.8 Field Day on TV (DCTV)

The Field Day on TV methodology consists of a programme of the Brazilian Agricultural

Research Corporation (Embrapa), linked to the Ministry of Agriculture, Livestock and Supply (MAPA), in which various themes are addressed, such as: agro-energy, agriculture and agro-industry, aquaculture, food security, environment, livestock, nanotechnology, germplasm banks, among many others. The main strategy of the programmes is to share technical information on highly relevant topics, which are previously selected based on proposals from all of Embrapa's units (BELTRÃO; PERREIRA, 2013).

The DCTV programmes are supported by various mass communication vehicles, particularly television channels, but are also available via the internet on the Embrapa portal (COSTA, 2014). Special care is taken in selecting the formats of the differentiated media and the language used, which should be the most appropriate to the predominant reality of family farmers (BELTRÃO; PEREIRA, 2013).

Costa (2014), in his work *"The case of the mangaba pickers"* observed that the pickers make use of a "didactic descriptive" speech, especially the methodological description, that used to share knowledge through the cultural baggage experienced by farmers, using universal language in the social context in which they are inserted.

If one considers only the history of the birth of the media, especially television, it may be that DCTV is not considered an innovative methodology for sharing technology and knowledge. However, considering that the considerable cultural aspects regarding the best efficiency of use, especially of younger people, it can be considered as a tool of wide reach and is characterized as innovative in the era of informationalization more adaptable to young people and even to those who little participate in technology sharing events.

The potentialities opened by new interactive technologies, such as the Internet and social networks offer a challenging and, at the same time, innovative reference towards breaking the unidirectionality and centrality of communications based on the mere transfer of information. In this context, DCTV is continuously concerned with the meaning of the messages it shares and seeks to measure the efficiency of its transmission (LIMA, 2004; BELTRÃO; PEREIRA, 2014).

2.9 Rural Prose

This is an Embrapa communication methodology that uses a radio programme specifically targeted at the family farming segment (COSTA, 2014) to share technical information and knowledge that contributes to improving the quality of life of people in the countryside, complemented with music by local artists, recipes and poetry (BELTRÃO; PEREIRA, 2013). The authors emphasise that "thousands of Brazilian homes receive the waves of Prosa Rural and learn about the low-cost and easily adopted technologies and products developed by Embrapa for young people and family farmers in the Brazilian semi-arid region, the Jequitinhonha Valley (MG) and the North, Midwest, Southeast and South regions of Brazil".

According to Beltrão and Pereira (2013), at the time the programme was already broadcast free of charge by a network of 1,167 commercial, university and community radios, at least once a week, on fixed days and times suitable to their programming schedules, in order to build loyalty among listeners to tune in always on the same days and times.

Prosa Rural is divided into 6 sections, as follows: *Um Dedo de Prosa* - a board intended for interviews; *Pitacos da Hora* - is intended for recipes or tips; *Favas Contadas* - focuses on poetry, songs and "causos"; *Fala, Produtor - the voice of experience - a* board intended for producers' testimonials; *Ao pé do ouvido* - is intended for services or citizenship tips; and *Um dedo de prosa ou Favas contadas - a* board intended for radio drama. These can have specific content according to each edition of the program (BELTRÃO; PEREIRA, 2013).

In 2011, in the Northeast region and the Jequitinhonha Valley, an audio entitled "Conservation of *mangaba trees*" was distributed for broadcasting. According to Costa (2014, p. 50), some **excerpts** stand out, but only those that are most relevant to the validation of this methodology are listed. **Technician**: (...) *we, technicians, visited mangaba gatherers all over Brazil to get initial information: who they are, where they live, what they do. As time went by, we felt the need to bring them together to talk among themselves, exchange experiences and list the main threats and demands.* **Researcher**: *we, the researchers, learned a lot about the types of mangaba fruit, among other things, and the women waste pickers about how to produce mangaba seedlings.* The author points out that, by observing the audio excerpts, one can see that there is a great interaction of the actors involved, making it possible to induce the citizen that is listening to build new knowledge and incorporate it into productive processes, based on the information received through the Prosa Rural programme.

Another important aspect to be considered, as it is the situation of thousands of farmers, according to Miura and Beltrão (2009, p. 19), citing Ferrareto (2001, p. 97), describes that the [...] Radio is the newspaper of those who cannot read, is the master of those who can go to school, is the free entertainment of the poor. The authors identified that there is still low interaction with rural extension and, consequently, little participation of extensionists in the planning and potentialisation of the programme to promote the appropriation of technologies in the states and municipalities.

Therefore, this methodology when considered by cultural aspects and, reflecting on its efficiency to reach the audiences most in need of information, as well as the need for better use of time in the information age, represents a challenging alternative among the forms of communication, in order to enhance the processes of sharing technologies and knowledge (BELTRÃO; PEREIRA, 2013).

2.10 Collective Intervention

This is an old methodology, but it is rarely used. It was used from 2007 to 2011 as part of the project "Pilot Nuclei for Information and Technological Management for Family Agriculture", which corresponds to component 3 of Agrofuture, a programme developed by Embrapa with funds from the Inter-American Development Bank and the Federal Government. The main objective was to develop "pilot" institutional arrangements in the territories of Grande Dourados (MS), Nordeste Paraense (PA) and Sisal (PE), as an alternative capable of catalysing the efforts of public, private and non-governmental organisations with a view to enhancing the technological and management development of family farming, contributing to its economic and social viability (NASCIMENTO et al., 2011).

Collective intervention" consists of interacting with the reality in which development takes place, with a focus on family farming, involving various actors, using its own dynamics, not from a pre-established model, but by building a regionalized local consensus at the municipal and regional levels, expanding the range of possibilities, generating new paths for the future of rural communities, from the adoption of appropriate technologies and building new knowledge (SILVA JUNIOR et al., 2013).

Nascimento (et al. 201 1) emphasize that the methodology's viability depends on all entities respecting each other's characteristics and mission. The authors also call attention to the fact that the collectivist sense must prevail over institutional vanities.

One of the first steps as part of the "collective intervention" is the participatory diagnosis of reality, identifying the main production chains, the major problems faced and the possible technological solutions to solve them, characterized as the "time-zero" in the timeline (PADOVAN et al., 2011b); PADOVAN et al., 2013). Then the actions involving the main productive chains are planned, involving collective activities of technology sharing, knowledge building, among others of interest to the collectivity (SILVA JUNIOR et al., 2013). In this context, the sharing of technologies, together with the construction of knowledge, are exercised in various ways and with various methodological tools (PADOVAN et al., 2011b).

As part of the collective intervention, according to Padovan et al. (2011b), priority is given to social technologies that have good potential for providing solutions to the main problems faced by farmers. This approach can also be seen as a method and technique for promoting the empowerment of farmers, collectively building development alternatives that stem from innovative experiences, enabling new forms of knowledge construction.

Another important aspect of this collective construction, with technologies, processes and knowledge as instruments of innovation, refers to the harmonization between technical-scientific and popular knowledge, built from generation to generation. In some situations, various traditional knowledge is applied in an innovative way, proving to be efficient, complementing the knowledge generated by research (PADOVAN et al., 2011b). The authors point out that by adding social technologies to traditional knowledge, there are significant improvements in processes, increased productivity and product quality, contributing to the well-being and better quality of life of the families involved.

Concluding remarks

There is a good diversity of innovative methodologies for sharing technology and knowledge. However, it is strategic to choose the one that is most appropriate for the type of audience you want to reach. Whenever possible, it is essential to involve the target audience in technology sharing actions, through collective construction, so that they are also protagonists of the process and not just objects of the actions.

The greatest innovations in the process of sharing technologies occur when different methodologies are adopted and that they complement each other so that the target audience has easy access to the technologies and the respective pertinent information, which facilitates their understanding of their essences and how to adopt and incorporate them into their production processes.

The different actors involved in agriculture and cattle-raising need to understand objectively the need to strengthen the "organizational networks" and act in an articulated manner to that end. These networks are powerful tools to boost the sharing of technologies and the construction of new knowledge that will benefit their members, rural properties, municipalities and regions.

Bibliographical references

ALVES, R. N. B.; MODESTO JÚNIOR, M. S. Roça sem fogo e trio da produtividade da mandioca. **Inclusão Social**, v. 6, n. 1, 2013.

BALBINO, L. C.; CORDEIRO, L. A. M.; PORFIRIO-DA-SILVA, SILVA, V. P.; MORAES, A.; MARTINEZ, G. B.; ALVARENGA, R. C.; KICHEL, A. N.; FONTANELI, R. S.; SANTOS, H. P.; FRANCHINI, J. C.; GALERANI, P. R. Evolução tecnológica e arranjos produtivos de sistemas de integração lavoura-pecuária-floresta no Brasil. **Pesquisa Agropecuária Brasileira**, v. 46, n. 10 pp. 0-0, 2011.

BALBINO, L. C.; PORFIRIO-DA-SILVA, V.; KICHEL, A. N.; ROSINHA, R. O.; DA COSTA, J. A. A. **Manual orientador para implantação de unidades de referência tecnológica-URT em integração lavoura-pecuária-floresta- iLPF**. Planaltina, DF: Embrapa Cerrados. 2011.

BALSADI, O. V.; CRUZ, M. C.; VERNE, M. C.; PEREIRA, V. F.; SICOLI, A. H. (Org.). **Technology transfer and knowledge construction**. Brasília, DF: Embrapa Informação Tecnológica, 2013.

BELTRÃO, S. L. Projeto Unaí: pesquisa e desenvolvimento em assentamentos de reforma agrária. **Cadernos de Ciência & Tecnologia**, v. 27, n. 1/3, p. 105-111, 2013.

BELTRÃO; S. L.; PEREIRA; F. A. Minibibliotecas, Prosa Rural e Dia de Campo na TV: communicative and pedagogical actions mediating dialogue for sustainable development. In: BALSADI, O. V.; CRUZ, M. C.; VERNE, M. C.; PEREIRA, V. F.; SICOLI, A. H. (Org.). **Technology transfer and construction of knowledge**. Brasília, DF: Embrapa Informação Tecnológica, 2013. p. 353-369.

BRANDÃO, C. R.; STRECK, D. R. A pesquisa participante e a partilha do saber: uma introdução. **Ideias e Letras**, Aparecida, SP, p. 7-20, 2006.

CEREDA, M. P.; VILPOUX, O. Metodologia para divulgação de tecnologia para agroindústrias rurais: exemplo do processamento de farinha de mandioca no Maranhão. **Revista Brasileira de Gestão e Desenvolvimento Regional**, Taubaté, SP, v. 6, n. 2, p. 119250, 2010.

COSTA, P. R.; BRAGA JUNIOR, S. S.; GALINA, S. V, R. Cooperação com Fontes Externas de Tecnologia: Estratégia e Gestão. In: **ENCONTRO NACIONAL DE PÓS- GRADUAÇÃO E PESQUISA EM ADMINISTRAÇÃO** - ENANPAD, 31, Rio de Janeiro, RJ. Annals... Rio de Janeiro, RJ, 2007.

COSTA, V. C. **Mulher e extrativismo na comunicação da pesquisa agropecuária - O caso das Catadoras de mangaba**. Dissertation (Master in Language Studies). State University of Campinas, Campinas, 2014.

DE ALMEIDA, R. G.; MACEDO, M. C. M.; ALVES, F. V. Integración de sistemas de cultivo-ganadería-forestal con énfasis en la producción de carne. In: CONGRESSO COLOMBIANO, 2; **SEMINAR INTERNACIONAL SILVOPASTOREO**, 1. Universidad Nacional de Colombia, Medellín, 2012.

DE NOVAES, M. B. C.; GIL, A. C. A pesquisa-ação participante como estratégia metodológica para o estudo do empreendedorismo social em administração de empresas. **Revista de Administração Mackenzie**, v. 10, n. 1, p. 134 - 160. 2009.

DOMIT, L. A. Adaptation of training and visit to Brazil. In: DOMIT, L. A.; LIMA, D. de; ADEGAS, F. S.; DALBOSCO, M.; GOMES, C.; OLIVEIRA, A. B.; CAMPANINI, S. M. S. (Orgs.). **Manual de implantação do Treino e Visita (T&V)**. Londrina, PR: Embrapa Soja, 2007. 86 p. (Embrapa Soja. Documentos, 288).

DOMIT, L. A. Capacitação continuada: uma alternativa para aprimorar o processo de transferência de tecnologias. In: BALSADI, O. V.; CRUZ, M. C.; VERNE, M. C.; PEREIRA, V. F.; SICOLI, A. H. (Org.). **Technology transfer and knowledge construction**. Brasília, DF: Embrapa Informação Tecnológica, 2013. p. 316-328.

ESCOBAR, J. L.; PEREIRA, F. A. Pesquisa de audiência e recepção do programa de rádio Prosa Rural: primeiros resultados. In: **ENCONTRO DOS GRUPOS DE RESQUISA EM COMUNICAÇÃO, 10; CONGRESSO BRASILEIRO DE CIÊNCIAS DA COMUNICAÇÃO**, 33. 2010. Caxias do Sul, RS. Proceedings, 2010. CD-ROM.

FONTES, M. A.; RABANAL, J. E. M.; FILHO, E. S. R. 'A Roça do futuro': the construction of the Peasant to Peasant methodology in the South of Sergipe. **Geonordeste**, v. 1, p. 1-26, 2013.

FRIGOTTO, G. **Política de formação continuada do servidor público**: uma alternativa metodológica à doutrina neoliberal. OFICINA CONCEPTUAL "CONSTRUINDO A GESTÃO PÚBLICA", v. 1, 2011.

GASTAL, M. L. Research with and for family farmers and traditional communities. In: BALSADI,

O. V.; CRUZ, M. C.; VERNE, M. C.; PEREIRA, V. F.; SICOLI, A. H. (Org.). **Technology transfer and construction of knowledge**. Brasília, DF: Embrapa Informação Tecnológica, 2013. p. 33-55.

GIMONET, J. C. Nascimento e desenvolvimento de um movimento educativo: as Casas Familiares Rurais de Educação e Orientação. In: **SEMINARIO INTERNACIONAL DA PEDAGOGIA DA ALTERNÂNCIA: ALTERNANCE AND DEVELOPMENT**, 1. 1999. Proceedings... Salvador: UNEFAB, 1999. p. 39-48.

HOLT GIMÉNEZ, E. **Campesino a campesino**: Voces de Latinoamérica Movimiento Campesino para la Agricultura Sustentable. Managua: SIMAS, 2008.

LIMA, V. A. **Mídia**: Teoria e política. 2 ed. São Paulo: Fundação Perseu Abramo, 2004.

LÓPEZ RODRIGUES, E. **Campesino a Campesino Nicaragua**: los principios del promotor voluntario. Managua: Unión Nacional Del Agricultores y Ganaderos, 2008.

MAGALHÃES, M. S. **Escola Família Agrícola**: uma escola em movimento. 2004. 126 p. Dissertation (Master's Degree in Agronomy). Federal University of Espírito Santo, Vitória, 2004.

MARTINS, M. V. F.; GALERANI, P. R. A metodologia treino e visita (T&V). In: DOMIT, L. A.; LIMA, D. de; ADEGAS, F. S.; DALBOSCO, M.; GOMES, C.; OLIVEIRA, A. B.; CAMPANINI, S. M. S. (Orgs.). **Manual de implantação do Treino e Visita (T&V)**. Londrina: Embrapa Soja, 2007. 86 p. (Embrapa Soja. Documentos, 288).

MIURA, J.; BELTRÃO, S. L. **Prosa Rural** - Manual de Produção e Edição. Brasília - DF: Embrapa Informação Tecnológica, 2009. 149 p.

NASCIMENTO, P. P.; SICOLI, A. H.; MARTINS, M. A. G.; BALSADI, O. V.; SILVA JÚNIOR, C. D. (Org.). **Innovations in territorial development**: New challenges for Embrapa. Brasília, DF: Embrapa Informação Tecnológica, 2011.

NICODEMO, M. L. F.; MELOTTO, A. M. 10 anos de pesquisa em sistemas agroflorestais em Mato Grosso do Sul. In: CONGRESSO SISTEMAS AGROFLORESTAIS E DESENVOLVIMENTO SUSTENTÁVEL. Campo Grande: Embrapa Gado de Corte, 2013.

OLIVEIRA, M. C. B.; LIMA, D. **A visão sobre transferência de tecnologia na Embrapa**. In: DOMIT, L. A.; LIMA, D. de; ADEGAS, F. S.; DALBOSCO, M.; GOMES, C.; OLIVEIRA, A. B.; CAMPANINI, S. M. S. (Orgs.). Manual de implantação do Treino e Visita (T&V). Londrina: Embrapa Soja, 2007. 86 p. (Embrapa Soja. Documentos, 288).

OLIVEIRA, M. N.; XAVIER, J. V.; ALMEIDA, S. C. R.; SCOPEL, E. **Projeto Unaí: pesquisa e desenvolvimento em assentamentos de reforma agrária**. Brasília, DF: Embrapa Informação Tecnológica, 2009. 264 p

PADOVAN, M. P.; KOMORI, O. M.; PADOVAN, D. S. S.; LEONEL, L. A. K. Unidades-Referência: uma experiência inovadora para validação e socialização de tecnologias e processos "com e para" os produtores rurais. In: **SEMINÁRIO INTERNACIONAL CAMPO, EDUCAÇÃO E DIVERSIDADE**, 1, 2011, Dourados, MS. **Annals**. Dourados: UFGD, 2011a. CD-ROM.

PADOVAN, M. P.; KOMORI, O. M.; ALMEIDA, A. C.; PADOVAN, D. S. S.; LEONEL, L. A. K. Informação e gestão tecnológica para a agricultura familiar no Território da Grande Dourados, MS: uma experiência inovadora em construção. In: NASCIMENTO, P. P.; SICOLI, A. H.; MARTINS, M. A. G.; BALSADI, O. V.; SILVA JÚNIOR, C. D. (Org.). **Innovations in territorial development**: New challenges for Embrapa. Brasília, DF: Embrapa Informação Tecnológica, 2011b. p. 165-201.

PADOVAN, M. P.; KOMORI, O. M.; PADOVAN, D. **S. S.**; PEZARICO, C. R.; LEONEL, L. A. K. Fortalecimento da agricultura familiar no Território da Grande Dourados, MS, a partir da gestão tecnológica, formação de capital social e valorização da produção. In: BALSADI, *O.* V.; CRUZ, M. C.; VERNE, M. C.; PEREIRA, V. F.; SICOLI, A. H. (Org.). **Technology transfer and the construction of knowledge**. Brasília, DF: Embrapa Informação Tecnológica, 2013. p. 95-122.

PASSADOR, C. S.; LOPES, J. E. F. Educação do campo no Estado de São Paulo: análise do nível de ruralidade das escolas no desempenho escolar. **Revista do Serviço Público**, v. 65, p. 87-113, 2014.

PERERA, A. R. F. **Avaliação da Rede de Referência como estratégia de transferência de tecnologia na perspectiva dos agricultores**. 2009. 94 f. Dissertation (Master in Family Agriculture) - Federal University of Pelotas, Pelotas, 2009.

PETERSEN, P.; SILVEIRA, L. Construção do conhecimento agroecológico em redes de agricultores-experimentadores: a experiência de assessoria ao Pólo Sindical da Borborema. In: **ENCONTRO NACIONAL DE AGROECOLOGIA**, 2, Recife, 2007. Cadernos... Recife: ANA, 2007. p.103-130.

PIRES, A. H. B.; SANTOS, J. A. Multiplicação de sistemas agroflorestais: a experiência do Centro Sabiá no agreste de Pernambuco. In: **ENCONTRO NACIONAL DE AGROECOLOGIA**, 2, Recife, 2007. Proceedings... Recife: ANA, 2007. p. 217-232.

PLONSKI, G. A. Bases for a movement for technological innovation in Brazil. **São Paulo em Perspectiva**, v. 19, n. 1, p. 25-33, 2005.

RAI, V.; SCHULTZ, K.; FUNKHOUSER, E. International low carbon technology transfer: Do intellectual property regimes matter? **Global Environmental Change** v. 24, p. 60-70, 2014.

RIGAMONTI, M. J. S.; SANTOS, J. A.; SOUZA, H. M. R. Alternância: uma educação para o desenvolvimento rural sustentável. In: BALSADI, O. V.; CRUZ, M. C.; VERNE, M. C.; PEREIRA, V. F.; SICOLI, A. H. (Org.). **Technology transfer and construction of knowledge**. Brasília, DF: Embrapa Informação Tecnológica, 2013. p. 305-314.

SANTOS, P. S. S. **Intervenientes do processo de transferência de tecnologias**: um estudo de caso na Embrapa. 2013. 147 p. Dissertation (Master in Administration). School of Economics and Finance - IBMEC.

SARKAR, S. **The innovative entrepreneur**: make it different and conquer your space in the market. Rio Janeiro: Elsevier, 2008.
SCANDOLARA, A.; MAROCCO, A.; PIETRI, V.; ROSSI, E.; MAZZONI, B.; BATTILANI, P. Management of *Fusarium verticillioides* in maize. **Journal of Plant Pathology**, v. 90, p. 325-326, 2008.

SILVA JUNIOR, J. F.; MACHADO, Y. M. C.; RODRIGUES FILHO, W. B.; CAVALCANTI FILHO, L. F. M.; CASTRO, M. F.; FIGUEIROA, J. G. **Território da Mata Sul Pernambucana**. In: NASCIMENTO, P. P.; SICOLI, A. H.; MARTINS, M. A. G.; BALSADI, O. V.; SILVA JÚNIOR, C. D. (Org.). **Innovations in territorial development**: New challenges for Embrapa. Brasília, DF: Embrapa Informação Tecnológica, 2011. p. 271-299.

SILVA JUNIOR, J. F.; MACHADO, Y. M. C.; RODRIGUES FILHO, W. B.; CASTRO, M. F.; FIGUEIROA, J. G.; CAVALCANTI FILHO, L. F. M. Experiências de gestão territorial no Núcleo-Piloto de Informação e Gestão Tecnológica para a Agricultura Familiar do Território da Mata Sul Pernambucana. In: BALSADI, O. V.; CRUZ, M. C.; VERNE, M. C.; PEREIRA, V. F.; SICOLI, A. H. (Org.). **Technology transfer and the construction of knowledge**. Brasília, DF: Embrapa

Informação Tecnológica, 2013. p. 169-191.

SILVA, M. A. S; SIQUEIRA, E. R.; MEDEIROS, S. S.; MANOS, M. G. L.; TEIXEIRA, O. A.; SANTOS, R. F.; ALMEIDA, M. R. M.; RODRIGUES, R. F. de A.; MORAES, A. da C.; SANTOS, A. V.; MATOS, L. N. Social modeling as a tool for analysis of conflicting demands in rural territories. In: **SIMPOSIO SOBRE INOVAÇÃO E CRIATIVIDADE CIENTÍFICA NA EMBRAPA,** 2, 2010. Brasília: Embrapa. CD-ROM.

SILVA, M. E.; CORRÊA, A. M.; GÓMEZ, C. P. Innovating for sustainable consumption: The challenge in building a new organizational paradigm. **Revista de Negócios,** v. 17, n. 2, p. 72-90, 2012.

SIQUEIRA, E. R., FONTES, M. A.; SIQUEIRA, P. Z. R.; RABANAL, J. E. M.; SOUZA, H. C. Metodologia Camponês a Camponês na difusão de sistemas agroflorestais sucessionais no Nordeste do Brasil. **CONGRESSO BRASILEIRO DE SISTEMAS AGROFLORESTAIS,** 9. Ilhéus, BA: Instituto Cabruca, 2013. CD-ROM.

SPIELMAN, D. J.; EKBOIR, J.; DAVIS, K. The art and science of innovation systems inquiry: applications to Sub-Saharan African agriculture. **Technology in Society,** v. 31, n. 4, p. 399-405, 2009.

TEIXEIRA, E. S.; BERNARTT, M. L.; TRINDADE, G. A. Estudos sobre Pedagogia da Alternância no Brasil: revisão de literatura e perspectivas para a pesquisa. **Educação e Pesquisa,** v. 34, n. 2, p. 227-242, 2008.

ZEN, A. C. **The influence of resources and competencies in innovation: A multiple case study in the electronics industry of Rio Grande do Sul.** 2007. 139 p. Dissertation (Master's in Administration) - Universidade Federal do Rio Grande do Sul, Porto Alegre, RS, Brazil

CHAPTER 2

SOCIO-ECONOMIC AND ENVIRONMENTAL IMPACTS OF GROWING GREEN MANURES PRIOR TO MAIZE UNDER AGRO-ECOLOGICAL MANAGEMENT

Summary

Green manure technology plays a strategic role in different cropping arrangements with crops of economic interest, resulting in significant improvements to the soil and agroecosystems as a whole. However, there is a great lack of studies involving socioeconomic and even environmental aspects. In this context, a research work was developed with family farmers, with the aim of identifying and describing the socioeconomic and environmental impacts arising from the adoption of the cultivation of spring/summer green manures prior to maize-tillage in production systems on an agro-ecological basis. The research was conducted from January to November 2014, involving producers from several municipalities in the state of Mato Grosso do Sul. Data were obtained through interviews with nineteen rural producers, users of the technology, which were carried out using a script according to AMBITEC-SOCIAL and AMBITEC-AGRO. A complementary evaluation was also carried out with the interviewees, by means of questions prepared by researchers from Embrapa Agropecuária Oeste. The results showed positive social impacts, highlighting the improvement in income generation of rural establishments, increase in the diversity of income sources and value of properties, improvement in environmental and personal health, in addition to significant increases in institutional relationships. The greatest environmental impacts refer to the drastic reduction in the use of agrochemicals/chemical inputs and/or materials, as well as a significant increase in the productive capacity of the soil. However, this technology is still little adopted, since there is little dissemination to farmers. They point out that there is a need to establish reference units on farms with green manures so that farmers can learn about the experiences of other farmers in rural communities through technical visits, field days and other events. They also emphasize the need to develop illustrated material, with accessible language for farmers, and to provide training for technical assistance.

Keywords: Green manure, economic viability, positive social impacts, environmental and personal health.

Abstract

Green manure technology develops strategic role in different arrangements of crops with cultivation of economic interest, resulting in significant improvements to the soil and agro- ecosystems as a whole. However, there is a great shortage of studies on socio-economic aspects involved and even the environment. In this context, developed a research work with family farmers, with the goal of identifying and describing the social, economic and environmental impacts arising from the adoption of cultivation of green manures to spring/summer preceding the off-season maize crop production systems in agroecological foundations. The survey was conducted from January to November 2014, involving producers of several municipalities in the State of Mato Grosso do Sul. The data were obtained through interviews with nineteen rural producers, users of technology, which were carried out using a script as the AMBITEC-AMBITEC AGRO-SOCIAL. An additional assessment was also carried out with respondents through questions drawn up by researchers at Embrapa Agropecuária Oeste. The results showed positive social impacts, including improving income generation of rural establishments, increased diversity of source of income and the value of the properties, environmental and personal health improvement, in addition to significant increases in institutional relationships. The biggest environmental impacts refer to the drastic reduction in the use of agrochemicals/chemical inputs and or materials, as well as a significant increase of the productive capacity of the soil. However, this technology is still not adopted, since there is little dissemination to farmers. Highlight that there is need for deployment of reference units in rural properties with green manures, so that producers can meet other experiences in rural communities from producers of technical visits, field days and other events. Also emphasize the need for preparation of material illustrated with language accessible to producers and conducting training for technical assistance.

Keywords: Green manure, economic feasibility, positive social impacts, environmental and personal health.

1 Introduction

Family farming has always been considered one of the pillars of economic dynamism in a significant part of developed countries, collaborating to help the distribution of wealth and promote the development of society (TOMAS et al., 2012).

In this direction, it would not be incorrect to state that the development of a country may be directly related to the development of its family agriculture, which can be a powerful tool to ensure food security for the world population and future generations, in addition to enabling the reduction of hunger and poverty in Brazil (FAO, 2012). However, it cannot be ignored that the implementation of the fertilizer industry in the country from the 1970s onwards discouraged the practice of green manuring even though excellent results were obtained in the most diverse production conditions (WUTKE, 1993), (MATEUS & WUTKE, 2006).

The need to develop family farming in a sustainable manner is currently under discussion, seeking low environmental impact and increased economic yield for small producers. Land use is sustainable, according to Lopes and Alves (2005), when productivity is adequate in the economic sphere, ecologically acceptable, socially and culturally viable. Still according to those authors, the lack of infrastructure in the production processes determines the sustainability of agricultural practice through the need for intensive land use. In this sense, family farmers have few means of increasing production, thus green manuring is a system of "conservationist production and high productivity" (LOPES, ALVES, 2005).

In the state of Mato Grosso do Sul, family farming has been gaining expressiveness, driven by public policies that in recent years have been intensified at the federal level, as described by Sangalli and Schilindwein (2012).

On the other hand, the idea that the world population is supported by a kind of "industrialized agriculture", with a high degree of specialization, less diversity and greater use of chemical products has been absorbed, justified by the economic viability defended by large companies in this sector. This process has subjected family farming to high vulnerabilities due to the risks of monoculture or monoactivity, the high production costs of these systems that are highly dependent on external inputs and the high contributions of resources needed to fund these systems (AUDEH et al., 2011).

In this context, it is of fundamental importance for the sustainable development of family farming to use alternative production systems that reduce the need for external inputs, that promote lower environmental impacts, that are more diversified and that enhance natural processes in agro-ecosystems. To this end, it is important to use each parcel of land according to its suitability, carrying capacity and economic productivity, so that natural resources are made available to man for their best use and benefits, while being conserved for future generations (LEAL et al., 2010).

Green manuring is an agricultural process that uses some plant species in planting systems that can be in rotation, succession or intercropping with the main crop. As part of this process, green manure technology can assume a strategic role in different cropping arrangements with crops of economic interest, resulting in significant improvements to the soil and agroecosystems as a whole (SAGRILO, 2009).

Green manuring is an age-old agricultural practice that aims to improve the productive capacity of soils by providing undecomposed organic plant material, which is produced by plants grown exclusively for this purpose, managed at the beginning of the reproductive cycle (CUNHA et al., 2008).

The species used as green manures are strategic in crop rotation systems and for intercropping with crops of economic interest, providing expressive benefits, such as: 1) rapid soil coverage and great production of mass for the soil system, being able to improve its level of organic matter; 2) good production of mass for formation of the mulch, favouring the no-till system; 3) recycling of nutrients leached in depth, i.e. recovery of nutrients that would be lost to the deeper soil layers; 4) supply of nitrogen fixed directly from the atmosphere by legumes; 5) intensification of biological activities in the soil; 6) increase in the storage capacity of water in the soil; 7) protection of the soil against wind, rain and solar radiation in a short period of time; 8) decrease in the infestation of weeds and in the incidence of pests and pathogens in crops of economic interest; 9) decompaction of the soil and improvement in the structuring and circulation of air in the soil; 10) decrease in the variation of soil temperature (more constant temperature); 11) improvement in the use and efficiency of fertilizers and correctives; 12) aid in the recovery of low fertility soils (PADOVAN et al., 2006).

To grow green manures before the crop of economic interest, one of the main recommended practices consists of growing grasses or other non-leguminous species, preceding the planting of the legume of food and/or commercial interest, using species such as oats, rye, turnip rape and millet, for example. The plants are grazed or cut with a knife-roller when the green manures are at the beginning of grain formation and then direct sowing of the legume crop, such as beans, soybeans, peas, among others, preferably without the use of herbicides (PADOVAN et al., 2013a).

On the other hand, when crops such as maize, rice, sunflower, sorghum, among others, that are not nitrogen-fixing or have low nitrogen-fixing capacity are to be planted, pre-cultivation of leguminous species is recommended, because studies have shown that legumes have the ability to provide, through biological fixation, all the nitrogen needed for crops of food and/or commercial interest (PADOVAN et al., 2013a; PADOVAN et al., 2013b).

The possibility of reducing the amount of nitrogen to be applied to the soil with chemical fertilizers is a relevant aspect in the economic impact of crops. Scivittaro et. al. (2000) tells us that compared with mineral fertilizers, the efficiency of green manures for crops is low, not exceeding 20% in the first crop after application, thus the combination of green manures and mineral fertilizers is for those authors the most viable management alternative to obtain both immediate and long-term effects with the use of mainly Mucuna Preta, however, for rice, i.e. green manuring offers nitrogen fixation results similar to those obtained with the use of mineral fertilizers.

Oliveira (2010) points out the use of Brachiária *brizantha* also as an alternative to legumes in intercropping with corn due to the formation of straw and accumulation of dry matter in the soil cover.

The following table illustrates legumes according to their nitrogen fixation capacity.

Table 1 Nitrogen fixation potential of legumes in green manure.

Common name	Amount of N fixed (kg/ha)
Guandu	37-280
Cowpeas	49-190
Crotálaria breviflora	98-160
Crotalaria juncea	150-450
Crotalaria mucronata	80-160
Crotalaria ochroneuca	133-200
Crotalaria spectabilis	60-120
Labelabe	66-180
Chicharo	80-100
White lupin	128-268
Black mucuna	120-210
Grey mucuna	170-210
Dwarf mucuna	50-100
Vetch	90-180

Source: adapted from Mateus & Wutke (2006)

Some of the legumes are easier to manage and produce seeds, enabling farmers to save money by buying seeds in the market.

Santos (et. al., 2007) point out that legumes are preferred when the intention is to form organic matter in the soil because of the large mass produced, their richness in mineral elements and their capacity to mobilize nutrients from the soil, besides taking advantage of nitrogen from the atmosphere. Amado (2000, *apud* Santos 2007) and confirmed by Gonçalves et. al. (2001, *apud* Santos, 2007) conclude that green manure is the indicated alternative to complement the supply of nitrogen and that it is capable of increasing productivity in consortium with mineral nitrogen. However, green manure occupies the area for a period, preventing the succession of two crops of economic value, which may eventually constitute an obstacle to the family farming system. One option may be intercropping, where the green manure can be sown simultaneously or followed by the main crop.

Heinrichs (et.al., 2004), in an experiment to measure the soil yield in the cultivation of maize intercropped with green manures, concluded, after two years of application, that the highest maize grain production was in the system intercropped with pork bean and Crotolaria spectabilis also showed good results, both sown 30 days after maize and consequently constituting a good option for the small farmer.

Leal (et. al., 2005) developed a research on the economic viability of crop rotation such as that of maize preceded by green manure in typical cerrado vegetation soil using four plant species in a no-till farming system (SPD) and costing mapping by the Total Operational Cost method. In SPD the straw and crop rotation form the basic species that sustain the productivity of the soil of the cerrado with tendencies to nutritional poverty and whose climate is particularly aggressive.

In order to ensure the economic survival of family farming and improve the socioeconomic potential of small rural producers, but also adopted by large producers, safrinha corn, also called second crop, planted from January to April, soon after the summer crop, is preferably cultivated with green manure to be able to re-establish the balance of the system (OLIVEIRA et. al., 2013).

Thus, it is understood why the green manuring technology is an important option to strengthen family farming, activate or improve the local development process, favouring the adoption of agroecological principles. In this perspective, the question that arises is: "Can the green manure technology in cultivation preceding the corn crop be able to provide socioeconomic and environmental gains to producers?"

In the pursuit of a robust answer to the question proposed in view of the relevance of this theme for Brazilian society, a research project was developed with family farmers in the state of Mato Grosso do Sul, aiming to identify and describe the socioeconomic and environmental impacts arising from the adoption of the cultivation of spring/summer green manures preceding the maize-chicken crop in agro-ecological production systems.

2 Methodology

The research was conducted between January and November 2014, in the central-southern region of the state of Mato Grosso do Sul. The data were obtained through interviews with rural producers, users of the technology, which were carried out using a script according to the System for Evaluating the Social Impact of Technological Innovation in Agriculture - AMBITEC-SOCIAL (RODRIGUES et al., 2005) and the System for Evaluating the Social Impact of Agricultural Technological Innovations - AMBITEC-AGRO (RODRIGUES et al., 2003), available in Appendix I.

The integral AMBITEC system employs a practical platform (MS-EXCEL), of simple and low cost execution, which can be applied to every technological and environmental universe of institutional insertion. The set of spreadsheets allows the consideration of several environmental and socioeconomic aspects of contribution of a given technological innovation, depending on the segment or size of the agribusiness in question (IRIAS et al., 2004).

In parallel to using the model proposed by AMBITEC, a complementary assessment was carried out with the interviewees by means of questions prepared by the team from the Prospection and Evaluation of Technologies Sector of Embrapa Agropecuária Oeste (Annex II), with the aim of better understanding the initiative to adopt the technology, aspects that need to be implemented, the farmer's decision making when inserting new practices into his production system, among others.

To locate and identify the farmers to be interviewed, the *Snowball Sam*pling methodology was used (BAYLEY, 1994), which consists of discovering possible key informants, who were represented by the Agency for Agrarian Development and Rural Extension (Agraer), the Mato Grosso do Sul Organic Producers Association (APOMS), municipal governments, non-governmental organizations, social movements, public research institutions and other farmers' organizations (unions, rural community associations and cooperatives).

Nineteen interviews were conducted with family farmers from 12 municipalities (Table 2).

Table 2 Producers interviewed in the state of Mato Grosso do Sul about the impacts of technology and their respective municipalities of origin.

Municipalities	State	Producer Family		Patronal Producer		Total
		Small	Medium	Great	Commercial	
Dourados	MS	3	0	0	0	3
Ponta Porã	MS	2	0	0	0	2
Naviraí	MS	1	0	0	0	1
Terenos	MS	1	0	0	0	1
Bodoquena	MS	1	0	0	0	1
Juti	MS	1	0	0	0	1
Ivinhema	MS	5	0	0	0	5
Glory of Dourados	MS	1	0	0	0	1
New World	MS	1	0	0	0	1
Campo Grande	MS	1	0	0	0	1
Amambai	MS	1	0	0	0	1
Tacuru	MS	1	0	0	0	1
Total		19	0	0	0	19

Source: Elaborated by the author

3 Results and discussion

Due to the complementary evaluation to the pre-established indexes and parameters in Ambitec, it was observed that there was a very positive response in the sense of enriching the information generated from the use of AMBITEC. This complementary information raised made it possible to infer more precisely and clearly some aspects pointed out by the methodology references.

Furthermore, it was important to have a qualitative assessment by the producer on issues such as adoption or discontinuation of technology use, demands for related technology transfer actions, as well as technical information that can serve to improve the technology being evaluated.

This feedback from producers is also important for the researcher in the form of demand, which can be useful for the process of updating/innovating processes both in research and in new technology transfer actions.

The following discussion points show some results of the changes implemented and which are positive in the sense that they open space for other intrinsic aspects of the production process.

3.1 Analysis of economic impacts

In the evaluation of the economic impact of the green manure technology preceding the corn crop under agro-ecological management, the indicator of increased productivity was used, since the increased production obtained with the corn crop without the anticipation of green manures was compared.

The economic dimension includes not only the formal economy, but also informal activities with the aim of increasing financial income. Seen this way, profit would be the main objective, having in our study agriculture as a source of income (GROOT, 2002; REIS, 2011). On the other hand, Silva (1995), assumes that economic sustainability can be achieved by the efficient allocation of resources and innovations implemented on current mechanisms of investment orientation.

To evaluate the economic impact, the Net Gain per hectare obtained by adopting the technology and the Regional Economic Benefit were estimated.

The economic benefits from the period of the beginning of the adoption of the technology until now were considered to compare the evolution of both the area of adoption and the benefits generated.

Net gains result from the cultivation of corn in succession to green manures. In 2010, the year the technology was launched, the yield per production unit was R$ 500.00 per hectare. An increase in value was observed, reaching R$ 660.00 per hectare in 2014, as shown in the following table.

Table 3 Evolution of Net Income per Productive Unit

Year	Unit of Measurement UM	Previous Income/UM (A)	Current Income/UM (B)	Unit Price R$/UM (C)	Additional Cost R$/UM (D)	Unit Gain R$/UM E=[(B-A)xC]-D
2010		3.500,00	5.500,00	0,250	-	500,00
2011		3.500,00	5.500,00	0,390	-	780,00
2012		3.500,00	5.500,00	0,380	-	760,00
2013		3.500,00	5.500,00	0,320	-	640,00
2014		3.500,00	5.500,00	0,330	-	660,00

Source: Elaborated by the author

Embrapa Agropecuária Oeste participated in the development of the technology at 80%. Thus, Embrapa's net gain with the technology in 2010 was R$400 per hectare, while in 2014 it reached R$528 (Table 3). The regional economic benefits (BER), in the year the technology was launched in Mato Grosso do Sul, reached R$ 200,000. From this date the technology expanded to other states, but this analysis encompasses the areas earmarked for the use of the technology in the state of Mato Grosso do Sul. Thus, in 2014, the BER reached R$ 264,000.00 (Table 3).

Table 4 Total Regional Economic Benefits achieved with the technology

Year	Embrapa's Participation % (F)	Embrapa's Net Gain R$/UM G=(ExF)	Adoption Area: Unit of Measurement - UM	Adoption Area: QuantxUM (H)	Economic Benefit I=(GxH)
2010	80%	400,00		500	200.000,00
2011	80%	624,00	ha	500	312.000,00
2012	80%	608,00		500	304.000,00
2013	80%	512,00		500	256.000,00
2014	80%	528,00		500	264.000,00

Source: Elaborated by the author

3.2 Social Impact Assessment

The cultivation of green manures before corn crops under agro-ecological management promotes differentiated impacts on job creation, income generation, health promotion, as well as on the management and administration of the Production Units. The results obtained in the evaluations carried out in Mato Grosso do Sul are presented in Table 5.

Table 5 Social impact index (Triple of sustainability) provided by the cultivation of green manures prior to corn cultivation, under agro-ecological bases, in the state of Mato Grosso do Sul, in 2014.

Types of impacts	Indicators	General Average
Social impact - employment aspect	Capacity Building	0,00
Social impact - employment aspect	Local skilled employment opportunity	0,04
Social impact - employment aspect	Job vacancies and worker status	0,02
Social impact - employment aspect	Quality of employment	0,00
Social impact - income aspect	Generation of income for the establishment	5,38
Social impact - income aspect	Diversity of income sources	2,15
Social impact - income aspect	Property value	3,13
Social impact - health aspect	Environmental and personal health	1,06
Social impact - health aspect	Occupational health and safety	0,07
Social impact - health aspect	Food safety	0,00
Social impact - management and administration aspect	Dedication and profile of the person in charge	1,75
Social impact - management and administration aspect	Condition of marketing	0,33
Social impact - management and administration aspect	Waste recycling	0,00
Social impact - management and administration aspect	Institutional relationship	3,88
Social impacts - general average......................	....	**0,98**

Source: Elaborated by the author

3.2.1 Employment aspect

The social impacts related to employment can be seen in Table 4. The adoption of the technology did not imply an increase in the training of employees and/or persons responsible for the activity. Although this does not represent a positive index in the sample, it reflects another situation:

the profile of the interviewees is characterized by family farmers who usually already use green manure as an option for soil coverage, pest control, animal feed, among others. The innovation, in this respect, is to implement its use with a view to improving the grain production system and increasing productivity. Therefore, the employment and training aspect did not show high indexes when evaluated, as it did not represent a quantitative change regarding the adoption of the technology, only a change in the form of implementation of the crop by the producer. As for the opportunity of qualified local employment, the indicator resulted in an impact equal to 0.04, meaning that all jobs, in general family jobs, were dedicated to activities directly linked to routine activities in the farm.

The change in the job offer is restricted to hiring temporary workers, who are only hired for a short period of time. Of the 19 interviewees, only two reported this situation, or 10.53%. On the other hand, in similar cases, family engagement is observed, where the producers' children take over or partially manage the control of the property. This indicator obtained a positive impact equal to 0.02.

The quality of employment was not changed, pointing out that the activity maintains the working conditions characteristic of family involvement in conducting activities on the properties. However, in several statements the quality related to the health of the workers involved in the activity was highlighted, since by using agro-ecological practices, the risk of contamination/contact with agrochemicals is almost non-existent or even non-existent.

3.2.2 Income aspect

The social impacts related to income can be seen in Table 4. The green manure technology prior to the cultivation of corn under agro-ecological management pointed out, as a positive index, a high change in the generation of income in rural establishments, resulting in an impact equal to 5.38. Another index found that represents great impact is the value of the property related to the aspect of conservation of natural resources, which reached 3.13. Next, the index of diversification of sources of agricultural and livestock income in the establishments obtained a value of 2.15. Moderate improvement was also observed in the security, distribution and amount of income.

As mentioned above, the interviewees see the practice of using green manure before crops of commercial interest as a possibility of diversifying activities on the properties, with increases in productivity, and with this, the possibility of negotiating better opportunities with the market that buys the produce. This increase in the income of the properties is indicated by the interviewees as a result of the use of technology. The indicator "value of the property", which relates to the conservation of natural resources, was highlighted in the interviewees' view as one of the main benefits generated by technology, a fact that suggests the concern of producers with the available resources, as well as highlighting the importance of having accessible Technologies, Practices and Processes - TPPs that enable agricultural production with quality and commitment to improving the environment.

The diversification of agricultural income sources by using green manure for sowing prior to the cultivation of other species was also highlighted as one of the benefits generated on the properties, as it allows producers to use the green manure species to produce seeds and animal feed, for example.

The adoption of technology did not imply an increase in investment in improvements, since one of the main changes in the planning of activities refers to the planting of crops.

3.2.3 Health aspect

The social impacts related to health can be seen in Table 4. The technology improved environmental and personal health, with a positive impact equal to 0.48. This is mainly credited to the reduction in the use of pesticides, considering the agro-ecological practices directly related to the use of the technology. Other benefits pointed out refer to the reduction in the emission of atmospheric pollutants, water pollutants and the generation of soil contaminants. In general, there were no changes related to access to sports and leisure with the adoption of the technology.

In relation to occupational health and safety, which portrays the worker's exposure to hazardous and unhealthy factors, the impact of this variable corresponds to 0.32, indicating that exposure to these factors remained practically unchanged or was reduced. With this, we can see the concern of producers to improve working conditions and worker health. Considering that the profile of the interviewees already adopts in part several agro-ecological practices, this indicator was not highlighted with intensity. However, it was evident in several speeches that the option to use technology is also linked to more sustainable production practices, including in this case the reduced use of external inputs, mainly agrochemicals.

As for food safety, no change was observed in this aspect either, as the technology itself did not promote this gain for the producer. Food security is one of the fundamental principles that ensures the sustainability of the family unit and, therefore, is present in all processes related to the quantity and nutritional quality of the food provided. In this sense, the value attributed to this index is quite high in the production context, but it is not evidenced by only one practice but by the set of processes involving agricultural activities in rural establishments.

3.2.4 Management and administration aspect

Table 4 shows the social impacts inherent to management and administration. The green manure technology prior to maize cultivation under agro-ecological management provided improvements in training for the activity and a moderate increase in the use of the accounting system. This index is related to the dedication and profile of the person in charge and obtained a positive impact of 1.75. The use of the accounting system was mentioned by some interviewees as a form of control and planning, with a view to increasing crop productivity and consequent sale of the surplus, as well as the inclusion of new crops of economic interest in the production system.

Inherent in the profile of family farming, particularly for producers who choose to work on an agro-ecological basis, the search for knowledge is essential to stay in business. In the dynamics of these production systems, continuous updating is important, whether through formal training models or even through the exchange of experiences among farmers themselves.

As for institutional relationships, the value of 0.33 represented an increase in relation to the use of technical assistance and adherence to the associativism/cooperativism. This fact may be related to the need for technical assistance in view of the use of new practices , as well as the need for commercialisation of the surplus production generated.

4 Analysis of the Results of Social Impacts provided by the technology

The General Social Impact Index of the technology "Green manuring prior to maize cropping under agro-ecological management" resulting from the interviews with the producers who adopt it reached the value of 0.98, on a scale varying from -15 to +15.

Income generation was the variable with the highest index (5.38), indicating that the technology brought improvements in all aspects related to income, especially in factors such as security, stability, improved distribution and amount of income. One of the most frequently mentioned was the increase in productivity of the crops planted in succession to the green manures.

Another important and outstanding variable was the "value of the property", with a positive impact of 3.13. This value is due to the benefits generated by the technology that promote improvements in the conservation of natural resources, one of the main aspects highlighted by the interviewees. Another aspect listed by the sample was the diversification of agricultural income sources on the property, with an index equivalent to 2.15.

The indices that did not score, that is, remained unchanged were: training, quality of employment and waste recycling. These items were not improved by the adoption of the technology itself, since the agro-ecological practices used by the producers interviewed already include these aspects in their routine activities on the farms. This allows us to reflect on the qualitative aspects of the research, which despite not listing these benefits, they stand out because they are already present in the production system as a whole and not only related to one practice, in this case the use of green manure.

5 Environmental Impact Assessment

The adoption of green manure prior to corn cultivation, cultivated on agro-ecological bases, exerts peculiar effects on different environmental aspects, which are presented in Table 6.

Table 6 Environmental impacts provided by the cultivation of green manures preceding corn crops, under agro-ecological bases, in the state of Mato Grosso do Sul, in 2014.

Types of impacts	Indicators	General Average
Environmental impact - Technological efficiency	Use of agrochemicals/chemical inputs or materials	e 6,39
Environmental impact - Technological efficiency	Energy use	0,13
Environmental impact - Technological efficiency	Use of natural resources	0,00
Environmental Impact - Environmental conservation	Atmosphere	0,55
Environmental Impact - Environmental conservation	Productive capacity of the soil	9,93
Environmental Impact - Environmental conservation	Water	0,28
Environmental Impact - Environmental conservation	Biodiversity	0,49
Environmental Impact - Environmental restoration	Environmental recovery	0,06
Environmental Impacts - general average	....	**2,03**

Source: Elaborated by the author

5.1. Scope of Technology

It is estimated that 500 hectares were used in 2014 in the area where green manure technology is being used prior to maize crops under agro-ecological management in Mato Grosso do Sul.

Technological Efficiency

Table 5 shows that the indicator "use of agrochemicals" presents a coefficient of 6.39, which indicates a highly positive impact involving a significant reduction in the variables frequency, variety of active ingredients and toxicity of the phytosanitary products used by the technology adopters. Furthermore, there is a noticeable reduction in the use of formulated fertilizers and micronutrients, contributing significantly to the environmental improvement provided by the technology. This fact is directly related to the characteristics that the technology incorporates, especially the use of agro-ecological practices and the fixation of nitrogen to the soil by green manures (FERNANDES et al., 2014).

As for the use of fossil fuels, the impact coefficient of 0.13 indicates that the technology presents a moderate decrease in the use of diesel required to conduct the activity. Only 15% of the interviewees related an increase in the use of fuels in relation to the practices developed previously. In this case, there is a correlation with the area used for planting crops on the property. The opening of new areas for planting using the technology was not indicated, as observed in the item "Use of natural resources", but rather the expansion of the area of cultivation of crops of commercial interest in succession to the planting of green manure species.

5.3 Environmental Conservation

The environmental impacts related to environmental conservation can be seen in Table 5. The impact coefficient of 0.55 for atmosphere indicates that the technology provides a decrease in the emission of greenhouse gases, particulate material/smoke, odors and noise generation. This index is observed and the empirical evidence reported by the technology users corroborates the statement that there are improvements in aspects related to air quality, greenhouse gas emissions and soil quality. In this sense, the producers state that the adoption of the technology with the incorporation of green manure in the production system has contributed with a great increase in soil quality, indicated by the impact coefficient of 9.93. This is due to the reduction of erosion, the loss of organic matter and nutrients and the reduction of soil compaction. In most cases, the use of green manure contributes significantly to soil decompaction, since the volume and depth of the roots help in this process. Some species are especially recommended for this purpose. However, the greatest aspect of improvement in soil quality, as mentioned by the interviewees, refers to the contribution of organic matter and nutrients to the soil, especially the nitrogen provided by these crops. In the view of these producers, this is the major benefit that has contributed to increased crop productivity and reduced costs, especially when compared to the traditional system that uses nitrogen fertilization.

The improvement in water quality is observed with less impact, by the reduction of turbidity and silting up of rivers and streams, since there is no water runoff into the springs. The index obtained for this variable was 0.28.

The adoption of the technology indicates a value of 0.49 for the biodiversity indicator. In this item, there is no indication of loss of native vegetation, loss of fauna corridors and loss of caboclo species/varieties. On the contrary, interviewees state that the technology promotes the diversity of beneficial insects (biological control) and maintains the production of native seeds used on small properties. The aspects related to the use of agro-ecological practices favoured this situation.

5.4 Environmental Recovery

The impact coefficient of 0.06 indicates that the technology did not interfere significantly regarding the environmental recovery aspect. This value, however, does not

portrays the real situation experienced by these producers, because as reported by the interviewees and taking into account the practices already used on the properties, there was no direct interference in the recovery of degraded ecosystems, legal reserve and permanent preservation areas. This is because these aspects are already being built by producers within a set of actions and practices, where the technology implemented contributes to this process. Thus, in the view of these producers, the technology itself does not act separately to obtain these benefits, but is an integral part of a sum of practices that has already been incorporated over time. Therefore, the environmental recovery of these areas has been occurring with the use of various techniques, where the use of green manure complements and adds new improvements to the production systems. For this reason, the impact index of the use of the technology was not as evident as expected, despite having in its context the improvement in soil quality as its main benefit.

6 Environmental Impact Index

The evaluation of the green manure technology prior to maize cultivation under agro-ecological management is highly positive from an environmental point of view, obtaining an index of 2.03, considering a variable scale of -15 to +15. This result indicates that the technology, besides being an economically viable alternative, provides diversification of activities on the property and contributes significantly to environmental sustainability. This concept is well evidenced by producers who adopt the technology, because the gains obtained in productivity and improvement of the productive environment are perceived in different aspects: increased productivity, improved soil conditions and crop diversification in the production system, including options of commercial interest.

Another important and outstanding variable was the decrease in the use of agrochemicals, with an index of 6.39. The producers' report highlights that green manuring contributes, as one of the agro-ecological practices already adopted, mainly in the preservation of natural enemies that are essential for biological control, as well as for soil cover, which minimizes weed infestation in cultivation areas, reducing the need for labour.

7 Integrated and comparative evaluation of the impacts generated

The evaluation of the technology "Green manuring prior to corn cropping under agro-ecological management" was positive, mainly in two aspects related by the interviewees: improvement in soil quality and increase in productivity/crop yields in relation to the previous cropping pattern.

The incorporation of green manures mainly favoured the increase in soil organic matter and nitrogen incorporated by the legumes, as well as soil coverage and, consequently, improvement in water infiltration into the soil, reduction in losses by evapotranspiration, reduction in weed infestation and soil biota. These were the main aspects mentioned and observed. In the soil's productive capacity variable, the index of 9.93 evidenced this situation. *"Green manuring is a simple and cheap way to recover soil fertility that contributes to the recovery of the chemical and physical properties of the soil... it contributes to less pollution of the soil and rivers, being beneficial from a social, economic and environmental point of view"* (quote from an interviewed producer).

The variable "income generation" of the establishment was the one that obtained the highest social index, 5.38. This was another aspect pointed out by the interviewees, especially with regard to the low cost of implementing and using the technology. The values themselves, related to increases in productivity could not be evaluated, since they are based on data that were not measured under the same conditions for all cases. However, producers reported that there was a significant increase in yields as a result of the use of green manure in the production system. Another highlight is the inclusion of crops of economic interest in the production system together with the cultivation of green manures, which added other possibilities that did not previously exist and, therefore, the generation of income was complemented.

"The practice of green manuring is very important because of the gains both in improving soil fertility and increasing agricultural production" (Quote from a producer interviewed).

When asked about the main benefits of the technology, in order of importance they were listed: improvement of soil quality as most important for 74% of the interviewees; in second place, the low cost (57%) and as third option the increase in corn crop productivity (52%).

When asked about the non-use of this technology, the interviewees listed the following arguments: 47% of the interviewees believe that other producers do not adopt the technology because they are unaware of the practice, its benefits and how to use it. Around 31% believe that there is a lack of technical guidance and 22% believe or know someone who does not use the technology due to the difficulty in managing green manures.

In this sense, the actions for sharing technologies focused on the theme are important, since 47% of respondents reported that the choice to adopt the technology was recommended by Embrapa and/or technical assistance. When asked about the negative points that lead producers not to adopt the technology, 50% of respondents related that there is a low knowledge of this practice, due to little dissemination to farmers. Another 50% credited the high cost of seeds and the difficulty of finding them easily on the market.

Another issue related to technology sharing actions, and identified in the interviewees' statements, refers to the need to implement reference units in rural properties with green manures, so that producers can learn about the experiences of other producers in rural communities through technical visits, field days and other events. Allied to this, the preparation of material with accessible language for producers and training for technical assistance show as needs of great relevance to help disseminate this practice and increase its adoption by producers.

8 Estimated cost of the technology

To estimate the costs, the costs of the research project that resulted in the recommendation of the technology, from the beginning of its implementation in 2007, were taken into account. Thus, Table 6 was adjusted with the information collected since the beginning of the research work with the green manure technology prior to maize cultivation under agro-ecological management.

Table 7 Estimated costs.

Year	Personnel Costs	Research Funding	Capital Depreciation	Administration Costs	Technology Transfer Costs	Total
2007	16.310	3.434	773	1.674	1.427	23.618
2008	24.248	3.381	686	1.400	1.505	31.220

2009	26.804	3.306	755	1.990	2.670	35.525
2010	23.040			1.600	1.555	26.195
2011	18.900			1.748	1.500	22.148
2012	29.833			1.930	1.300	33.063
2013	31.539			1.940	6.800	40.279
2014	34.016			1.768	3.650	39.434
Total	**204.690**	**10.121**	**2.214**	**14.050**	**20.407**	**251.482**

Source: Elaborated by the author

8.1 Cost Analysis

The research work that helped develop this technology began in 2007, at Embrapa Agropecuária Oeste, with the aim of making the practice of using green manures before crops of food and economic interest increasingly viable. The corn crop stood out in this aspect, for obtaining significant productivity increments in response to the technology and for presenting differentials in its use in rural properties, from animal production, as well as in the commercialization of surpluses and food security.

Considering, for the purpose of this analysis the year 2007, the beginning of the evaluation of research efforts with green manures in agro-ecological based system, the cost of personnel, research costs, depreciation of capital, administration costs and technology transfer costs, until the year 2014, corresponds to R$ 251,482.00. Of this amount, R$10,121.00 was spent directly on research costs, corresponding to 4.02% of the total.

8.2 Cost/Benefit Analysis

The analysis of investments made with the technology considered an 8-year horizon. It was found that the technology is highly advantageous in all indicators evaluated (Tables 8 and 9).

Table 8 Analysis of technology investments.

Year	Flow of benefits	Cost flow	Net benefit flow	Internal Rate of Return	Cost/Benefit Ratio
Year 1	0	23.618	-23.618		
Year 2	0	31.220	-31.220		
Year 3	0	35.525	-35.525		
Year 4	200.000,00	26.195	173.805		
Year 5	312.000,00	22.148	289.852		
Year 6	304.000,00	33.063	270.937		
Year 7	256.000,00	40.279	215.721		
Year 8	264.000,00	39.434	224.566	6,05	2,16

Source: Elaborated by the author

The return on investment measured by the Internal Rate of Return (IRR), which represents the discount rate that equals the sum of cash flows to the value of the investment, was high, reaching 6.05%. This indicator indicates that the investments are economically viable, since they exceed the minimum rate of attractiveness.

The Benefit/Cost ratio was obtained by dividing revenues and the current value of costs. Thus, the analysis shows that the technology obtained an index of 2.16, indicating that the technology is efficient.

Considering the Minimum Rates of Attractiveness (TMA) of 4.0%, 6.0%, 8.0%, 10.0%, 12.0%, 14.0%, 16.0% and 18.0%, the Net Present Value (NPV), which corresponds to the sum of the expected cash flows brought to year zero, obtained by the difference between the present value of cash inflows and the present value of cash outflows, at discount rates mentioned, ranged from R$ 846.000.00 thousand, when the AAR was 4.0% to R$ 380,000.00, when the AAR was 18.0%. These results indicate that the amount of cash that the producer will have available at the end of the project is much higher than the investment made (Table 9).

Table 9 Net Present Value Analysis (in thousand BRL).

4%	6%	8%	10%	12%	14%	16%	18%
R$846	R$750	R$666	R$593	R$529	R$473	R$424	R$380

Source: Elaborated by the author

Final considerations

The viability of using green manure cultivation prior to maize is evidenced by family producers in Mato Grosso do Sul in several aspects:

1) Environmental - as a result of improved soil quality, reduced use of pesticides and growing concern over the use of natural resources;

2) Economic - due to the reduction in production costs with the continued use of the technology, less use of external inputs and increased productivity of corn grown after the green manures;

3) Social - through improved income generation and working conditions as a result of the drastic reduction in the use of agrochemicals, increased diversity of income sources and property value, improved environmental and personal health, as well as significant increases in institutional relationships.

The technology is still little adopted by farmers, since there is little dissemination of its benefits. The farmers highlight the need to establish reference units on farms with green manures so that farmers can learn about the experiences of other farmers in rural communities through interactive technical visits, field days and other collective activities. They also emphasize the need to develop illustrated technical material, with accessible language for producers, and to provide training for technical assistance teams.

Bibliographical references

AUDEH, S. J. S.; LIMA, A. C. R.; CARDOSO, I. M.; CASALINHO, H. D.; JUCKSCH, I. Soil quality: an ethnopedological view in family farms, producers of organic tobacco. **Revista Brasileira de Agroecologia**, v. 6, p. 34-48, 2011.

CUNHA, K. A. A.; MARASCA, R. V.; PADOVAN, M. P. Nível de adoção de boas práticas em sistemas de produção sob transição agroecológica no Sul do Mato Grosso do Sul. **Journal Brasileira de Agroecologia**, v. 3, p. 157-160, 2008.

FAO - Food and Agriculture Organization of the United Nations. FAO representation in Brazil. **The population increase and the challenges of food security**. FAO debates global production and demand for food in the Sebrae Knowledge Forum. Brasília, 2012. Available at: https://www.fao.org.br/apdsa.asp. Accessed on: 05 Sep. 2014.

FERNANDES, S. S. L.; MATOS, A. T.; MOITINHO, M. R.; MOTTA, I. S.; OTSUBO, A. A.; PADOVAN, M. P. Desempenho de adubos verdes num sistema de produção sob bases ecológicas em Itaquiraí, Mato Grosso do Sul. **Cadernos de Agroecologia**, v. 9, p. 1-12, 2014.

HEINRICHS, R., VITTI, G. C., FIGUEIREDO, P.A.M. Soil chemical attributes and corn grain yields under intercropping with green manures. **Revista Científica Eletrónica de Agronomia**. Ano III, n. 5, jun 2004.

LEAL, A. J. F.; LAZARINI, E., TARSITANO, M. A. A., SÁ, M. E. D., JUNIOR, F. G. G. Viabilidade económica da rotação de culturas e adubos verdes antecedendo o cultivo do milho em sistema de plantio direto em solo de cerrado. **Revista Brasileira de Milho e Sorgo**, v. 4, n. 3, 2010.

MATEUS,G.P., WUTKE, E.B. Espécies leguminosas utilizadas como adubos verdes. **Revista Pesquisa & Tecnologia**, vol. 3, n. 1, jan-jun 2006.

NETO, A. S.; MACIEL, L. S. B.; LAPOLLI, E. M.. The teacher and the educational proposals of the ratio studiorum: some initial reflections on teaching practice. **EDUCERE** (Mérida), v. 16, p. 273-281, 2012.

OLIVEIRA, P. **Consórcio de milho com adubos verdes e manejo da adubação nitrogenada no cultivo de feijão em sucessão no sistema integração Lavoura-Pecuária no Cerrado**. Tese (Doutorado), Escola Superior de Agricultura "Luiz de Queiroz" - ESALQ, 2010.

OLIVEIRA, L.S., SILVA, A.A.S., PAIVA, M.J.A., SILVEIRA, W.R. Avaliação da produtividade do milho safrinha em sucessão a adubos verdes no plantio. **Anais do XII Seminário Nacional do Milho Safrinha**. Dourados, 2013.

PADOVAN, M. P.; OLIVEIRA, F. L. de; CESAR, M. N. Z. O papel estratégico da adubação verde no manejo agroecológico do solo. In: PADOVAN, M. P. **Conversão de sistemas de produção convencionais para agroecológicos**: novos rumos à agricultura familiar. Dourados, 2006. p. 69-83.

PADOVAN, M. P.; MOTTA, I. S.; CARNEIRO, L. F.; MOITINHO, M. R.; SALOMAO, G. B.; RECALDE, K. M. G. Pre-cropping of green manures to corn in agroecosystem subjected to ecological management in the South of Mato Grosso do Sul. **Revista Brasileira de Agroecologia**, v. 8, p. 3-11, 2013a.

PADOVAN, M. P.; MOTTA, I. S.; CARNEIRO, L. F.; MOITINHO, M. R.; NASCIMENTO, J. S.; SALOMAO, G. B. Performance of green manures and minimum cultivation of cassava subjected to ecological management in a distroferric Red Latosol in Dourados, Mato Grosso do Sul. **Cadernos de Agroecologia**, v. 8, p. 1-5, 2013b.

RODRIGUES, G. S.; CAMPANHOLA, C., KITAMURA, P. C. **Avaliação de Impacto Ambiental da Inovação Tecnológica Agropecuária**: AMBITEC-AGRO. Jaguariúna: Embrapa Meio Ambiente. 95 p. 2003 (Embrapa Meio Ambiente. Documents, 34).

RODRIGUES, G. S.; CAMPANHOLA, C.; KITAMURA, P. C. et al. **Sistema de Avaliação de Impacto Social de Inovações Tecnológicas Agropecuárias** - Ambitec-Social. Jaguariúna: Embrapa Meio Ambiente, 2005 (Embrapa Meio Ambiente. Research and Development Bulletin).

SAGRILO, E., LEITE, L. F. C., DA SILVA GALVÃO, S. R., LIMA, E. F. **Manejo agroecológico do solo**: os benefícios da adubação verde. Embrapa Meio-Norte, 2009.

SANGALLI, A. R.; SCHLINDWEIN, M. M. A contribuição da agricultura familiar para o desenvolvimento rural de Mato Grosso do Sul. **Redes**, v. 18, p. 82-99, 2013.

SCIVITTARO, W.B., MURAOKA, T., BOARETTO, A. E., TRIVELIN, P. C. Utilizacao de nitrogen de adubos verdes e mineral pelo milho. **Revista Brasileira de Ciência do Solo**, vol. 24, n. 4, 2000. PP. 917-926.

TOMAS, R. N.; SPROESSER, R. L.; BATALHA, M. O. Convenções, Capital Social e Desenvolvimento Efetivo na Agricultura familiar: O caso de Mato Grosso do Sul. **Organizações Rurais & Agroindustriais**, v. 14, n. 3, 2012.

WUTKE, E.B. Adubação verde: manejo da fitomassa e espécies utilizadas no Estado de São Paulo. In: WUTKE, E.B.; BULISANI, E.A.; MASCARENHAS, H.A.A. (Coords.) CURSO **SOBRE ADUBAÇÃO VERDE NO INSTITUTO AGRONÔMICO,** 1. 1993, Campinas: Instituto Agronômico,1993. p.17-29.

ANNEXES

ANNEX 1

Interview schedule of AMBITEC-SOCIAL and AMBITEC-AGRO
Environmental Dimension

Technological Efficiency

Tabela de coeficientes de alteração da variável			Pesticidas			Fertilizantes			Averiguação fatores de ponderação
Uso de Agroquímicos			Freqüência	Variedade de ingredientes ativos	Toxicidade	NPK hidrossolúvel	Calagem	Micronutrientes	
Fatores de ponderação k			-0,2	-0,2	-0,3	-0,1	-0,1	-0,1	-1
Escala máxima = pontual	Sem efeito	Marcar com X							
	Pontual	5							
	Local	-							
	Entorno	-							
Coeficiente de impacto = (coeficientes de alteração * fatores de ponderação)			0	0	0	0	0	0	0,0

Tabela de coeficientes de alteração da variável

Uso de Energia			Combustíveis fósseis				Biomassa				Eletricidade	Averiguação fatores de ponderação
			Óleo combustível / Carvão mineral	Diesel	Gasolina	Gás	Álcool	Lenha / Carvão vegetal	Bagaço de cana	Restos vegetais		
Fatores de ponderação k			-0,1	-0,1	-0,1	-0,1	-0,075	-0,075	-0,075	-0,075	-0,3	-1
Escala máxima = pontual	Sem efeito	Marcar com X										
	Pontual	5										
	Local	-										
	Entorno	-										
Coeficiente de impacto = (coeficientes de alteração * fatores de ponderação)			0	0	0	0	0	0	0	0	0	0,0

Tabela de coeficientes de alteração da variável

Uso de Recursos Naturais			Recurso natural			Averiguação fatores de ponderação
			Água para irrigação	Água para processamento	Solo para plantio (área)	
Fatores de ponderação k			-0,3	-0,3	-0,4	-1
Escala máxima = pontual	Sem efeito	Marcar com X				
	Pontual	5				
	Local	-				
	Entorno	-				
Coeficiente de impacto = (coeficientes de alteração * fatores de ponderação)			0	0	0	0,0

Environmental Conservation

Tabela de coeficientes de alteração da variável

Atmosfera			Tipo do poluente				Averiguação fatores de ponderação
			Gases de efeito estufa	Material particulado / Fumaça	Odores	Ruídos	
Fatores de ponderação k			-0,4	-0,4	-0,1	-0,1	-1
Escala da ocorrência =	Sem efeito	Marcar com X					
	Pontual	1					
	Local	2					
	Entorno	5					
Coeficiente de impacto = (coeficientes de alteração * fatores de ponderação)			0	0	0	0	0,0

Tabela de coeficientes de alteração da variável

Qualidade do Solo			Variável de qualidade do solo				Averiguação fatores de ponderação
			Erosão	Perda de matéria orgânica	Perda de nutrientes	Compactação	
Fatores de ponderação k			-0,25	-0,25	-0,25	-0,25	-1
Escala máxima = pontual	Sem efeito	Marcar com X					
	Pontual	5					
	Local	-					
	Entorno	-					
Coeficiente de impacto = (coeficientes de alteração * fatores de ponderação)			0	0	0	0	0,0

Tabela de coeficientes de alteração da variável					
Qualidade da Água	Variável de qualidade da água				Averiguação fatores de ponderação
	Demanda Bioquímica de Oxigênio	Turbidez	Espuma / Óleo / Materiais flutuantes	Sedimento / Assoreamento	
Fatores de ponderação k	-0,25	-0,25	-0,25	-0,25	**-1**
Escala da ocorrência =	Sem efeito	Marcar com X				
	Pontual	1				
	Local	2				
	Entorno	5				
Coeficiente de impacto = (coeficientes de alteração * fatores de ponderação)	0	0	0	0	**0,0**

Tabela de coeficientes de alteração da variável				
Biodiversidade	Variável de biodiversidade			Averiguação fatores de ponderação
	Perda de vegetação nativa	Perda de corredores de fauna	Perda de espécies / Variedades caboclas	
Fatores de ponderação k	-0,4	-0,3	-0,3	**-1**
Escala da ocorrência =	Sem efeito	Marcar com X		
	Pontual	1		
	Local	2		
	Entorno	5		
Coeficiente de impacto = (coeficientes de alteração * fatores de ponderação)	0	0	0	**0,0**

Environmental Recovery

Tabela de coeficientes de alteração da variável					
Recuperação Ambiental	Variável de recuperação ambiental				Averiguação fatores de ponderação
	Solos degradados	Ecossistemas degradados	Áreas de Preservação Permanente	Reserva Legal	
Fatores de ponderação k	0,2	0,2	0,2	0,4	**1**
Escala da ocorrência =	Sem efeito	Marcar com X				
	Pontual	1				
	Local	2				
	Entorno	5				
Coeficiente de Impacto = (coeficientes de alteração * fatores de ponderação)	0	0	0	0	**0,0**

Social Dimension

Aspect Employment

Tabela de coeficientes de alteração na capacitação

Capacitação	Tipo de capacitação			Nível da capacitação			Averiguação fatores de ponderação
	Local de curta duração	Especialização de curta duração	Oficial regular	Básico	Técnico	Superior	
Fatores de ponderação k	0,25	0,25	0,2	0,1	0,1	0,1	1
Escala da ocorrência = Sem efeito — Marcar com X							
Pontual 1	0	0	0	0	0	0	
Local 2							
Entorno 5							
Coeficiente de impacto = (coeficientes de alteração * fatores de ponderação)	0	0	0	0	0	0	0

Tabela de coeficientes de alteração da geração de emprego

Oportunidade de emprego local qualificado	Origem do trabalhador				Qualificação para a atividade				Averiguação fatores de ponderação
	Propriedade	Local	Município	Região	Braçal	Braçal especializado	Técnico médio	Técnico superior	
Fatores de ponderação k	0,25	0,2	0,15	0,1	0,025	0,05	0,1	0,125	1
Escala da ocorrência = Sem efeito — Marcar com X									
Pontual 1	0	0	0	0	0	0	0	0	
Local 2									
Entorno 5									
Coeficiente de impacto = (coeficientes de alteração * fatores de ponderação)	0	0	0	0	0	0	0	0	0

Tabela de coeficientes de alteração da oferta de emprego

Oferta de emprego e condição do trabalhador	Condição do trabalhador				Averiguação fatores de ponderação
	Temporário	Permanente	Parceiro / Meeiro	Familiar	
Fatores de ponderação k	0,1	0,2	0,35	0,35	1
Escala da ocorrência = Sem efeito — Marcar com X					
Pontual 1	0	0	0	0	
Local 2					
Entorno 5					

Tabela de coeficientes de alteração da qualidade do emprego									
Qualidade do emprego	Legislação trabalhista				Benefícios				Averiguação fatores de ponderação
	Prevenção do trabalho infantil	Jornada de trabalho <44h	Registro	Contribuição previdenciária	Auxílio moradia	Auxílio alimentação	Auxílio transporte	Auxílio saúde	
Fatores de ponderação k	0,2	0,2	0,2	0,2	0,05	0,05	0,05	0,05	1
Sem efeito / Marcar com X	x								
Escala da ocorrência = Pontual 1	0	0	0	0	0	0	0		
Local 2									
Entorno 5									
Coeficiente de impacto = (coeficientes de alteração * fatores de ponderação)	0	0	0	0	0	0	0	0	0

Income Aspect

Tabela de coeficientes de alteração na geração de renda					
Geração de renda	Atributos da renda				Averiguação fatores de ponderação
	Segurança	Estabilidade	Distribuição	Montante	
Fatores de ponderação k	0,25	0,25	0,25	0,25	1
Sem efeito / Marcar com X					
Escala da ocorrência = Pontual 1					
Local 2					
Entorno 5					
Coeficiente de impacto = (coeficientes de alteração * fatores de ponderação)	0	0	0	0	0

Tabela de coeficientes de alteração da diversidade de fontes de renda						
Diversidade de fontes de renda	Variável de diversificação de fontes de renda					Averiguação fatores de ponderação
	Agropecuária no estabelecimento	Não agropecuária no estabelecimento	Oportunidade de trabalho fora do estabelecimento	Ramificação empresarial	Aplicações financeiras	
Fatores de ponderação k	0,25	0,25	0,15	0,2	0,15	1
Sem efeito / Marcar com X						
Escala da ocorrência = Pontual 1						
Local 2						
Entorno 5						
Coeficiente de impacto (obs) = (coeficientes de alteração * fatores de ponderação)	0	0	0	0	0	0

Tabela de coeficientes de alteração do valor da propriedade						
Valor da propriedade	Variável de valor da propriedade					Averiguação fatores de ponderação
	Investimento em benfeitorias	Conservação dos recursos naturais	Preços de produtos e serviços	Conformidade c/legislação	Infraestrutura/ política tributária/ etc.	
Fatores de ponderação k	0,25	0,25	0,2	0,15	0,15	1
Sem efeito / Marcar com X						
Escala da ocorrência = Pontual 1						
Local 2						
Entorno 5						
Coeficiente de impacto = (coeficientes de alteração * fatores de ponderação)	0	0	0	0	0	0

Health Aspect

Tabela de coeficientes de alteração da saúde						
Saúde ambiental e pessoal	Variável de saúde ambiental e pessoal					Averiguação fatores de ponderação
	Focos de vetores de doenças endêmicas	Emissão de poluentes atmosféricos	Emissão de poluentes hídricos	Geração de contaminantes do solo	Dificuldade de acesso a esporte e lazer	
Fatores de ponderação k	-0,2	-0,2	-0,2	-0,2	-0,2	-1
Escala da ocorrência =	Sem efeito (Marcar com X)					
Pontual 1						
Local 2						
Entorno 5						
Coeficiente de impacto = (coeficientes de alteração * fatores de ponderação)	0	0	0	0	0	0

Tabela de coeficientes de alteração da variável de segurança ocupacional								
Segurança e saúde ocupacional	Exposição a periculosidade e fatores de insalubridade							Averiguação fatores de ponderação
	Periculosidade	Ruído	Vibração	Calor / Frio	Umidade	Agentes químicos	Agentes biológicos	
Fatores de ponderação k	-0,2	-0,1	-0,1	-0,1	-0,1	-0,2	-0,2	-1
Escala da ocorrência = Sem efeito (Marcar com X)								
Pontual 1								
Local 2								
Entorno 5								
Coeficiente de impacto = (coeficientes de alteração * fatores de ponderação)	0	0	0	0	0	0	0	0

Tabela de coeficientes de alteração da segurança alimentar				
Segurança alimentar	Variável de segurança alimentar			Averiguação fatores de ponderação
	Garantia da produção	Quantidade de alimento	Qualidade nutricional do alimento	
Fatores de ponderação k	0,3	0,3	0,4	1
Escala da ocorrência = Sem efeito (Marcar com X)				
Pontual 1				
Local 2				
Entorno 5				
Coeficiente de impacto = (coeficientes de alteração * fatores de ponderação)	0	0	0	0

Management and Administration Aspect

Tabela de coeficientes de alteração de variáveis de dedicação e perfil do responsável

Dedicação e perfil do responsável	Variável de dedicação do responsável						Averiguação fatores de ponderação
	Capacitação dirigida à atividade	Horas de permanência no estabelecimento	Engajamento familiar	Uso de sistema contábil	Modelo formal de planejamento	Sistema de certificação	
Fatores de ponderação k	0,2	0,2	0,15	0,15	0,15	0,15	1
Escala da ocorrência = Sem efeito (Marcar com X)							
Pontual 1							
Local 2							
Entorno 5							
Coeficiente de impacto = (coeficientes de alteração * fatores de ponderação)	0	0	0	0	0	0	0

Tabela de coeficientes de alteração da condição de comercialização

Condição de comercialização	Variável de comercialização							Averiguação fatores de ponderação
	Venda direta/ antecipada/ cooperada	Processamento local	Armazena-mento local	Transporte próprio	Propaganda/ Marca própria	Encadeamento com produtos/ atividades/ serviços anteriores	Cooperação com outros produtores locais	
Fatores de ponderação k	0,15	0,15	0,15	0,15	0,15	0,15	0,1	1
Escala da ocorrência = Sem efeito (Marcar com X)								
Pontual 1								
Local 2								
Entorno 5								
Coeficiente de impacto = (coeficientes de alteração * fatores de ponderação)	0	0	0	0	0	0	0	0

Tabela de coeficientes de alteração das medidas de reciclagem de resíduos

Reciclagem de resíduos	Variável de tratamento de resíduos domésticos			Variável de tramento de resíduos da produção		Averiguação fatores de ponderação
	Coleta seletiva	Compostagem/ reaproveitamento	Disposição sanitária	Reaprovei-tamento	Destinação ou tratamento final	
Fatores de ponderação k	0,2	0,2	0,2	0,2	0,2	1
Escala da ocorrência = Sem efeito (Marcar com X)						
Pontual 1						
Local 2						
Entorno 5						
Coeficiente de impacto = (coeficientes de alteração * fatores de ponderação)	0	0	0	0	0	0

Tabela de coeficientes de alteração de relacionamento institucional							
Relacionamento institucional	Variável de alcance institucional				Variável de capacitação contínua		Averiguação fatores de ponderação
	Utilização de assistência técnica	Associativismo/ Cooperativismo	Filiação tecnológica nominal	Utilização de assessoria legal/ vistoria	Gerente	Empregados especializados	
Fatores de ponderação k	0,2	0,2	0,15	0,15	0,15	0,15	1
Sem efeito — Marcar com X							
Escala da ocorrência = Pontual 1							
Local 2							
Entorno 5							
Coeficiente de impacto = (coeficientes de alteração * fatores de ponderação)	0	0	0	0	0	0	0

ANNEX 2

Interview schedule complementary to Ambitec, prepared by the team from the Technology Assessment and Prospection Sector of Embrapa Agropecuária Oeste

1 Why does the farmer make or has made the choice to use green manure?
( ) Seeking an increase in income with the crop planted in the sequence (maize, beans, another crop)
( ) Stability in production (maintain and improve productivity of maize, beans, other crop)
( ) Concern with natural resources (soil, water, air)
( ) Desire to try out a new technology
( ) Other reasons
Which ones: ____

2 What makes the farmer decide to use green manure on his property?
( ) By having seen the use of technology by other producers
( ) Technical assistance recommendation
( ) Self-initiative or curiosity
( ) Recommendations from research institutions, such as EMBRAPA, for example
( ) Other reasons
Which ones: ____

3 Why don't many farmers use green manure?
( ) Unawareness of the benefits of this practice
( ) Lack of technical guidance
( ) Operational difficulty in green manure management
( ) Little dissemination of green manure to farmers
( ) Other reasons
Which ones: ____

4 What is the main advantage perceived with the use of green manure by the farmer compared to

the management previously used? <u>It can be more than one option</u>
( ) Ease of planting and management of green manures
( ) Decrease in production costs
( ) Soil improvement
( ) Increase in productivity of crops planted after green manures
( ) Other advantages
Which ones: ____

5   What difficulties do farmers perceive in adopting green manuring?
( ) Difficulty in green manure management
( ) Increased cost/manpower
( ) Difficulty in planting species of commercial interest, such as maize and beans
( ) Difficulty in finding seeds
( ) High seed costs
( ) Fall in productivity of crops planted after green manures
( ) Other difficulties
Which ____

6   Those who did not adopt - what led the farmer not to adopt green manuring (Negative points).
( ) Difficult for farmers to harvest two crops in a year
( ) Lack of seeds on the market
( ) Increases the cost of production, as the seeds are expensive
( ) No advantage in green manure
( ) Low knowledge of this practice, due to little dissemination
( ) Other reasons.
Which ____

7   What could be done to favour the adoption of green manure by farmers - number from most important (1) to least important (7).
( ) Setting up green manure reference units in rural communities for visits by farmers
( ) Implementation of demonstrative units with green manure at Embrapa for farmer visits
( ) Carry out field days and technical visits to the Reference Units with green manuring in the rural communities
( ) Carry out field days and technical visits to the demonstration units with green manure in Embrapa Agropecuária Oeste
( ) Offer courses on green manure to farmers and extension technicians
( ) Publish illustrated material on green manure (booklet type)
( ) Stimulate the production of green manure seeds and the formation of community seed banks

8   Write, in a few words, what you think of green manure.

I want morebooks!

Buy your books fast and straightforward online - at one of world's fastest growing online book stores! Environmentally sound due to Print-on-Demand technologies.

Buy your books online at
www.morebooks.shop

Kaufen Sie Ihre Bücher schnell und unkompliziert online – auf einer der am schnellsten wachsenden Buchhandelsplattformen weltweit! Dank Print-On-Demand umwelt- und ressourcenschonend produziert.

Bücher schneller online kaufen
www.morebooks.shop

Printed by Books on Demand GmbH, Norderstedt / Germany